Introduction to Integral Calculus Systematic Studies with Engineering Applications

Introduction to Integral Calculus Systematic Studies with Engineering Applications

Editor

Jai Rathod

Introduction to Integral Calculus Systematic Studies with Engineering Applications

Edited by **Jai Rathod**

Printed in 2017

ISBN: 978-1-68117-186-9

Library of Congress Control Number: 2015949131

© 2016 by
SCITUS Academics LLC,
616, Corporate Way, Suite 2, 4766,
Valley Cottage, NY 10989

www.scitusacademics.com

Preface

An integral is a mathematical object that can be interpreted as an area or a generalization of area. Integrals, together with derivatives, are the fundamental objects of calculus. Other words for integral include antiderivative and primitive. The Riemann integral is the simplest integral definition and the only one usually encountered in physics and elementary calculus.

The study of integral calculus includes: integrals and their inverse, differentials, derivatives, anti-derivatives, and approximating the area of curvilinear regions.

Integration is an important function of calculus, and introduction to integral calculus combines fundamental concepts with scientific problems to develop intuition and skills for solving mathematical problems related to engineering and the physical sciences. The book provides a solid introduction to integral calculus and feature applications of integration, solutions of differential equations, and evaluation methods. This book explores the integral calculus and its plentiful applications in engineering and the physical sciences. A basic understanding of integral calculus combined with scientific problems, and throughout, the book covers the numerous applications of calculus as well as presents the topic as a deep, rich, intellectual achievement. The needed fundamental information is presented in addition to plentiful references.

Table of Contents

Calculus of Variations with Fractional Derivatives and Fractional Integrals

Ricardo Almeida and
Delfim F.M. Torres
Department of Mathematics,
University of Aveiro, 3810-193 Aveiro,
Portugal

ABSTRACT

We prove the Euler–Lagrange fractional equations and the sufficient optimality conditions for problems of the calculus of variations with functionals containing both fractional derivatives and fractional integrals in the sense of Riemann–Liouville.

INTRODUCTION

In recent years numerous works have been dedicated to the fractional calculus of variations. Most of them deal with Riemann–Liouville fractional derivatives (see [1], [2], [3], [4] and [5] and the references therein), a few with Caputo or Riesz derivatives [6], [7], [8] and [9]. Depending on the type of the functional being considered, different fractional Euler–Lagrange type equations are obtained. We also mention [10], where a fractional Euler–Lagrange equation is obtained corresponding to a prescribed fractional space. Here we propose a new kind of functional with a Lagrangian containing not only a Riemann–Liouville fractional derivative (RLFD) but also a Riemann–Liouville fractional integral (RLFI). We prove the necessary conditions of Euler–

Lagrange type for the fundamental fractional problem of the calculus of variations and for the fractional isoperimetric problem. Sufficient optimality conditions are also obtained under appropriate convexity assumptions.

FRACTIONAL CALCULUS

Fractional calculus is an interdisciplinary area, with many applications in several fields, such as engineering [11], [12] and [13], chemistry [14] and [15], electrical and electromechanical systems [16] and [17], viscoplasticity [18], physics [19], etc. [20].

In this section we review the necessary definitions and facts from fractional calculus. For more on the subject we refer the reader to [21], [22], [23] and [24].

Let f be a function defined on the interval [a,b]. Let α be a positive real and n:= [α] +1.

Definition 2.1

The left RLFI is defined by

$$_a I_x^\alpha f(x) = \frac{1}{\Gamma(\alpha)} \int_a^x (x-t)^{\alpha-1} f(t) dt,$$

and the right RLFI by

$$_x I_b^\alpha f(x) = \frac{1}{\Gamma(\alpha)} \int_x^b (t-x)^{\alpha-1} f(t) dt.$$

Calculus of Variations with Fractional Derivatives

The left RLFD is defined by

$$_aD_x^\alpha f(x) = \frac{d^n}{dx^n}\,_aI_x^{n-\alpha}f(x) = \frac{1}{\Gamma(n-\alpha)}\frac{d^n}{dx^n}\int_a^x (x-t)^{n-\alpha-1}f(t)dt,$$

while the right RLFD is given by

$$_xD_b^\alpha f(x) = (-1)^n\frac{d^n}{dx^n}\,_xI_b^{n-\alpha}f(x) = \frac{(-1)^n}{\Gamma(n-\alpha)}\frac{d^n}{dx^n}\int_x^b (t-x)^{n-\alpha-1}f(t)dt.$$

The operators of Definition 2.1 are obviously linear. We now present the rules of fractional integration by parts for RLFI and RLFD. Let $p \geq 1$, $q \geq 1$, and $1/p + 1/q \leq 1+\alpha$. If $g \in L_p(a,b)$ and $f \in L_q(a,b)$, then

$$\int_a^b g(x)\,_aI_x^\alpha f(x)dx = \int_a^b f(x)\,_xI_b^\alpha g(x)dx;$$

if f, g, and the fractional derivatives $_aD_x^\alpha g$ and $_xD_b^\alpha f$ are continuous on $[a,b]$, then

$$\int_a^b g(x)\,_aD_x^\alpha f(x)dx = \int_a^b f(x)\,_xD_b^\alpha g(x)dx, \quad 0 < \alpha < 1.$$

Remark 1

The left RLFD of f is infinite at $x=a$ if $f(a) \neq 0$ (cf. [25]). Similarly, the right RLFD is infinite if $f(b) \neq 0$. Thus, assuming that f possesses continuous left and right RLFD on $[a, b]$, then $f(a)=f(b)=0$ must be satisfied.

THE EULER–LAGRANGE EQUATION

Let us consider the following problem:

$$\mathcal{J}(y) = \int_a^b L(x, {}_aI_x^{1-\alpha}y(x), {}_aD_x^\beta y(x))dx \longrightarrow \min. \tag{1}$$

We assume that $L(\cdot,\cdot,\cdot) \in C^1([a,b] \times \mathbb{R}^2; \mathbb{R})$, $x \to \partial_2 L(x, {}_aI_x^{1-\alpha}y(x), {}_aD_x^\beta y(x))$ has continuous right RLFI of order $1-\alpha$ and $x \to \partial_3 L(x, {}_aI_x^{1-\alpha}y(x), {}_aD_x^\beta y(x))$ has continuous right RLFD of order β, where α and β are real numbers in the interval $(0,1)$.

Remark 2

We are assuming that the admissible functions y are such that ${}_aI_x^{1-\alpha}y(x)$ and ${}_aD_x^\beta y(x)$ exist on the closed interval $[a,b]$. We also note that as α and β goes to 1 our fractional functional $\mathcal{J}$ tends to the classical functional $\int_a^b L(x,y(x),y'(x))dx$ of the calculus of variations.

Remark 3

We consider functionals $\mathcal{J}$ containing the left RLFI and the left RLFD only. This comprise the important cases in applications. The results of the paper are easily generalized for functionals containing also the right RLFI and/or right RLFD.

Theorem 3.1: The Fractional Euler–Lagrange Equation

Let $y\,(\cdot)$ be a local minimizer of problem (1). Then, $y(\cdot)$ satisfies the fractional Euler–Lagrange equation

$$_x I_b^{1-\alpha} \partial_2 L(x, {}_a I_x^{1-\alpha} y(x), {}_a D_x^\beta y(x)) + {}_x D_b^\beta \partial_3 L(x, {}_a I_x^{1-\alpha} y(x), {}_a D_x^\beta y(x)) = 0 \quad \text{for all } x \in [a, b]. \tag{2}$$

Remark 4

Condition (2) is only necessary for an extremum. The question of sufficient conditions for an extremum is considered in Section 6.

Proof

Since y is an extremizer of $\mathcal{I}$, by a well-known result of the calculus of variations the first variation of $\mathcal{I}(\cdot)$ is zero at y, i.e.,

$$0 = \delta \mathcal{I}(\eta, y) = \int_a^b ({}_a I_x^{1-\alpha} \eta \, \partial_2 L + {}_a D_x^\beta \eta \, \partial_3 L) dx. \tag{3}$$

Integrating by parts,

$$\int_a^b {}_a I_x^{1-\alpha} \eta \, \partial_2 L dx = \int_a^b \eta \, {}_x I_b^{1-\alpha} \partial_2 L dx \tag{4}$$

and

$$\int_a^b {}_a D_x^\beta \eta \, \partial_3 L dx = \int_a^b \eta \, {}_x D_b^\beta \partial_3 L dx. \tag{5}$$

Substituting (4) and (5) into Eq. (3), we find that $\int_a^b \left({}_x I_b^{1-\alpha} \partial_2 L + {}_x D_b^\beta \partial_3 L \right) \eta dx = 0$ for each η. Since η is an arbitrary function, by the fundamental lemma of the calculus of variations we deduce that $_x I_b^{1-\alpha} \partial_2 L + {}_x D_b^\beta \partial_3 L = 0 \cdot$

Remark 5

As α and β goes to 1, the fractional Euler–Lagrange equations (2) becomes the classical Euler–Lagrange equation $\partial_2 L - d/dx \partial_3 L = 0$.

A curve that is a solution of the fractional differential equation (2) will be called an extremal of $\mathcal{J}$. Extremals play also an important role in the solution of the fractional isoperimetric problem (see Section 5). We note that Eq. (2) contains right RLFI and right RLFD, which are not present in the formulation of problem (1).

SOME GENERALIZATIONS

We now give some generalizations of Theorem 3.1.

Extension to Variational Problems of Non-commensurate Order

We now consider problems of the calculus of variations with Riemann–Liouville derivatives and integrals of non-commensurate order, i.e., we consider functionals containing RLFI and RLFD of different fractional orders. Let

$$\mathcal{J}(y) = \int_a^b L(x, {}_aI_x^{1-\alpha_1}y(x), \ldots, {}_aI_x^{1-\alpha_n}y(x), {}_aD_x^{\beta_1}y(x), \ldots, {}_aD_x^{\beta_m}y(x))dx, \qquad (6)$$

where n and m are two positive integers and $\alpha_i, \beta_j \in (0,1)$, i=1,...,n and j=1,...,m. Following the proof of Theorem 3.1, we deduce the following result.

Theorem 4.1

If $y(\cdot)$ is a local minimizer of (6), then $y(\cdot)$ satisfies the Euler–Lagrange equation

$$\sum_{i=1}^{n} {}_xI_b^{1-\alpha_i}\partial_{i+1}L + \sum_{j=1}^{m} {}_xD_b^{\beta_j}\partial_{j+n+1}L = 0 \quad \text{for all } x \in [a, b].$$

Extension to Several Dependent Variables

We now study the case of multiple unknown functions $y_1,\ldots,y_n$.

Theorem 4.2

Let $\mathcal{J}$ be the functional given by the expression

$$\mathcal{J}(y_1, \ldots, y_n) = \int_a^b L(x, {}_aI_x^{1-\alpha}y_1(x), \ldots, {}_aI_x^{1-\alpha}y_n(x), {}_aD_x^{\beta}y_1(x), \ldots, {}_aD_x^{\beta}y_n(x))dx.$$

If $y_1(\cdot),\ldots,y_n(\cdot)$ is a local minimizer of $\mathcal{J}$, then it satisfies for all $x \in [a,b]$ the following system of n fractional differential equations:

$${}_xI_b^{1-\alpha}\partial_{k+1}L + {}_xD_b^{\beta}\partial_{n+k+1}L = 0, \quad k = 1, \ldots, n.$$

Proof

Denote by y and η the vectors $(y_1,\ldots,y_n)$ and $(\eta_1,\ldots,\eta_n)$, respectively. For a parameter ϵ, we consider a new function

$$J(\epsilon) = \mathcal{J}(y + \epsilon\eta).$$

(7)

Since $y_1(\cdot),\ldots,y_n(\cdot)$ is an extremizer of $\mathcal{I}$, $J'(0)=0$. Differentiating Eq. (7) with respect to ϵ, at $\epsilon=0$, we obtain

$$\int_a^b \left[{}_aI_x^{1-\alpha}\eta_1\,\partial_2 L + \cdots + {}_aI_x^{1-\alpha}\eta_n\,\partial_{n+1}L + {}_aD_x^\beta\eta_1\,\partial_{n+2}L + \cdots + {}_aD_x^\beta\eta_n\,\partial_{2n+1}L \right] dx = 0.$$

Integrating by parts leads to

$$\int_a^b \left[{}_xI_b^{1-\alpha}\partial_2 L + {}_xD_b^\beta\partial_{n+2}L \right]\eta_1 + \cdots + \left[{}_xI_b^{1-\alpha}\partial_{n+1}L + {}_xD_b^\beta\partial_{2n+1}L \right]\eta_n\,dx = 0.$$

Consider a variation $\eta = (\eta_1, 0,\ldots, 0)$, η_1 arbitrary; then by the fundamental lemma of the calculus of variations we obtain ${}_xI_b^{1-\alpha}\partial_2 L + {}_xD_b^\beta\,\partial_{n+2}L=0$. Selecting appropriate variations η, one deduce the remaining formulas.

THE FRACTIONAL ISOPERIMETRIC PROBLEM

We consider now the problem of minimizing the functional $\mathcal{I}$ given by (1) subject to an integral constraint $I(y)=\int_a^b g\big(x, {}_aI_x^{1-\alpha}y(x), {}_aD_x^\beta\, y(x)\big)dx = I$, where I is a prescribed value. This problem was solved in [8] for functionals containing Caputo fractional derivatives and RLFI. Using similar techniques as the ones discussed in [8], one proves the following:

Theorem 5.1

Consider the problem of minimizing the functional $\mathcal{I}$ as in (1) on the set of functions y satisfying condition $I(y) = I$. Let y be a local minimum for the problem. Then, there exist two constants λ_0 and λ, not both zero, such that y satisfies the Euler–Lagrange equation ${}_xI_b^{1-\alpha}\partial_2 K + {}_xD_b^\beta\partial_3 K=0$ for all $x \in [a,b]$, where $K=\lambda_0 L+\lambda g$.

Remark 6

If y is not an extremal for I, then one can choose $\lambda_0 = 1$ in Theorem 5.1: there exists a constant λ such that y satisfies $_x I_b^{1-\alpha} \partial_2 F + _x D_b^\beta \partial_3 F = 0$ for all $x \in [a,b]$, where $F = L + \lambda g$.

SUFFICIENT CONDITIONS

In this section we prove the sufficient conditions that ensure the existence of minimums. Similarly to what happens in the classical calculus of variations, some conditions of convexity are in order.

Definition 6.1

Given a function L, we say that $L(x, u, v)$ is convex in $S \subseteq \mathbb{R}^3$ if $\partial_2 L$ and $\partial_3 L$ exist and are continuous and verify the following condition:

$$L(x, u + u_1, v + v_1) - L(x, u, v) \geq \partial_2 L(x, u, v)u_1 + \partial_3 L(x, u, v)v_1$$

for all $(x,u,v),(x,u+u_1,v+v_1) \in S$.

Similarly, we define convexity for $L(x, u, v)$.

Theorem 6.2

Let $L(x, u, v)$ be a convex function in $[a,b] \times \mathbb{R}^2$ and let y_0 be a curve satisfying the fractional Euler–Lagrange equation (2). Then, y_0 minimizes (1).

Proof

The following holds:

$$\mathcal{J}(y_0 + \eta) - \mathcal{J}(y_0) = \int_a^b \left[L(x, {_a}I_x^{1-\alpha} y_0(x) + {_a}I_x^{1-\alpha}\eta(x), {_a}D_x^\beta y_0(x) + {_a}D_x^\beta \eta(x)) - L(x, {_a}I_x^{1-\alpha} y_0(x), {_a}D_x^\beta y_0(x)) \right] dx$$

$$\geq \int_a^b \left[\partial_2 L(x, {_a}I_x^{1-\alpha} y_0(x), {_a}D_x^\beta y_0(x)) {_a}I_x^{1-\alpha}\eta + \partial_3 L(x, {_a}I_x^{1-\alpha} y_0(x), {_a}D_x^\beta y_0(x)) {_a}D_x^\beta \eta \right] dx$$

$$= \int_a^b \left[{_x}I_b^{1-\alpha} \partial_2 L + {_x}D_b^\beta \partial_3 L \right]_{(x, {_a}I_x^{1-\alpha} y_0(x), {_a}D_x^\beta y_0(x))} \eta \, dx = 0.$$

Thus, $\mathcal{J}(y_0+\eta) \geq \mathcal{J}(y_0)$.

We now present a sufficient condition for convex Lagrangians on the third variable only. First we recall the notion of exact field.

Definition 6.3

Let $D \subseteq \mathbb{R}^2$ and let $\Phi: D \rightarrow \mathbb{R}$ be a function of class C^1. We say that Φ is an exact field for L covering D if there exists a function $S \in C^1(D, \mathbb{R})$ such that

$$\partial_1 S(x, y) = L(x, y, \Phi(x, y)) - \partial_3 L(x, y, \Phi(x, y))\Phi(x, y),$$
$$\partial_2 S(x, y) = \partial_3 L(x, y, \Phi(x, y)).$$

Remark 7

This definition is motivated by the classical Euler–Lagrange equation. Indeed, every solution $y_0 \in C^2 [a,b]$ of the differential equation $y' = \Phi(x, y(x))$ satisfies the (classical) Euler–Lagrange equation $\partial_2 L - \dfrac{d}{dx}\partial_3 L = 0$.

Calculus of Variations with Fractional Derivatives

Theorem 6.4

Let $L(x,u,v)$ be a convex function in $[a,b] \times \mathbb{R}^2$, Φ an exact field for L covering $[a,b] \times \mathbb{R} \subseteq D$, and y_0 a solution of the fractional equation

$$_aD_x^\alpha y(x) = \Phi(x,\, _aI_x^{1-\alpha}y(x)). \tag{8}$$

Then, y_0 is a minimizer for $\mathcal{I}(y) = \int_a^b L\left(x,_aI_x^{1-\alpha}y(x), {}_aD_x^\alpha y(x)\right)dx$ subject to the constraint

$$\{y : [a, b] \to \mathbb{R} \mid {}_aI_a^{1-\alpha}y(a) = {}_aI_a^{1-\alpha}y_0(a),\, {}_aI_b^{1-\alpha}y(b) = {}_aI_b^{1-\alpha}y_0(b)\} \tag{9}$$

Proof

Let $E(x,y,z,w) = L(x,y,w) - L(x,y,z) - \partial_3 L(x,y,z)(w-z)$. First observe that

$$\frac{d}{dx}S(x,\, _aI_x^{1-\alpha}y(x)) = \partial_1 S(x,\, _aI_x^{1-\alpha}y(x)) + \partial_2 S(x,\, _aI_x^{1-\alpha}y(x))\frac{d}{dx}{}_aI_x^{1-\alpha}y(x)$$
$$= \partial_1 S(x,\, _aI_x^{1-\alpha}y(x)) + \partial_2 S(x,\, _aI_x^{1-\alpha}y(x))_aD_x^\alpha y(x).$$

Since $E \geq 0$, it follows that

$$\mathcal{I}(y) = \int_a^b \left[E(x,\, _aI_x^{1-\alpha}y,\, \Phi(x,\, _aI_x^{1-\alpha}y),\, _aD_x^\alpha y) + L(x,\, _aI_x^{1-\alpha}y,\, \Phi(x,\, _aI_x^{1-\alpha}y)) \right.$$
$$\left. + \partial_3 L(x,\, _aI_x^{1-\alpha}y,\, \Phi(x,\, _aI_x^{1-\alpha}y))(_aD_x^\alpha y - \Phi(x,\, _aI_x^{1-\alpha}y))\right]dx$$
$$\geq \int_a^b \left[L(x,\, _aI_x^{1-\alpha}y,\, \Phi(x,\, _aI_x^{1-\alpha}y)) + \partial_3 L(x,\, _aI_x^{1-\alpha}y,\, \Phi(x,\, _aI_x^{1-\alpha}y))(_aD_x^\alpha y - \Phi(x,\, _aI_x^{1-\alpha}y))\right]dx$$
$$= \int_a^b \left[\partial_1 S(x,\, _aI_x^{1-\alpha}y) + \partial_2 S(x,\, _aI_x^{1-\alpha}y)_aD_x^\alpha y\right]dx$$
$$= \int_a^b \frac{d}{dx}S(x,\, _aI_x^{1-\alpha}y)dx$$
$$= S(b,\, _aI_b^{1-\alpha}y(b)) - S(a,\, _aI_a^{1-\alpha}y(a)).$$

Because y_0 is a solution of (8), $E\left(x_{,a}I_x^{1-\alpha}y_0, \phi\left(x_{,a}I_x^{1-\alpha}y_0\right), aD_x^\alpha y_0\right)=0$. With similar calculations as before, one has

$$\mathcal{J}(y_0) = S(b, {}_aI_b^{1-\alpha}y_0(b)) - S(a, {}_aI_a^{1-\alpha}y_0(a)).$$

We just proved that $\mathcal{J}(y_0) \leq \mathcal{J}(y)$ when subject to the constraint (9).

CONCLUSIONS

In this note we consider a new class of fractional functionals of the calculus of variations that depend not only on fractional derivatives but also on fractional integrals. We exhibit necessary and sufficient conditions of optimality for the fundamental problem of the calculus of variations and for problems subject to integral constrains (isoperimetric problems). As future work it would be interesting to address the question of existence of solutions, and to study direct methods to minimize the proposed type of functionals.

ACKNOWLEDGMENTS

The authors are grateful to the support of the Control Theory Group from the Centre for Research on Optimization and Control given by the Portuguese Foundation for Science and Technology (FCT), cofinanced through the European Community Fund FEDER/POCI 2010.

REFERENCES

1. T.M. Atanacković, S. Konjik, S. Pilipović, Variational problems with fractional derivatives: Euler Lagrange equations, J. Phys. A: Math. Theor. 41 (9) (2008) 095201.

2. D. Baleanu, S.I. Muslih, E.M. Rabei, On fractional Euler–Lagrange and Hamilton equations and the fractional generalization of total time derivative, Nonlinear Dynam. 53 (1–2) (2008) 67–74.

3. R.A. El-Nabulsi, D.F.M. Torres, Necessary optimality conditions for fractional action-like integrals of variational calculus with Riemann–Liouville derivatives of order (α, β), Math. Methods Appl. Sci. 30 (15) (2007) 1931–1939.

4. G.S.F. Frederico, D.F.M. Torres, A formulation of Noether's theorem for fractional problems of the calculus of variations, J. Math. Anal. Appl. 334 (2) (2007) 834–846.

5. G.S.F. Frederico, D.F.M. Torres, Fractional conservation laws in optimal control theory, Nonlinear Dynam. 53 (3) (2008) 215–222.

6. O.P. Agrawal, Fractional variational calculus and the transversality conditions, J. Phys. A: Math. Gen. 39 (2006) 10375–10384.

7. O.P. Agrawal, Fractional variational calculus in terms of Riesz fractional derivatives, J. Phys. A 40 (24) (2007) 6287–6303.

8. R. Almeida, D.F.M. Torres, Necessary and sufficient conditions for the fractional calculus of variations with Caputo derivatives (submitted for publication).

9. D. Baleanu, Om.P. Agrawal, Fractional Hamilton formalism within Caputo's derivative, Czechoslovak J. Phys. 56 (10–11) (2006) 1087–1092.

10. S.I. Muslih, D. Baleanu, Fractional Euler–Lagrange equations of motion in fractional space, J. Vib. Control 13 (9–10) (2007) 1209–1216.

11. N. Ferreira, F. Duarte, M. Lima, M. Marcos, J.T. Machado, Application of fractional calculus in the dynamical analysis and control of mechanical manipulators, Fract. Calc. Appl. Anal. 11 (1) (2008) 91–113.

12. V.V. Kulish, J.L. Lage, Application of fractional calculus to fluid mechanics, J. Fluids Eng. 124 (3) (2002) 803–806.

13. R. Magin, Fractional calculus in Bioengineering. Part 1–3, Critical Reviews in Bioengineering 32 (2004).

14. R. Hilfer, Fractional diffusion based on Riemann–Liouville fractional derivatives, J. Phys. Chem. B 104 (16) (2000) 3914–3917.

15. F. Metzler, W. Schick, H.G. Kilian, T.F. Nonnenmacher, Relaxation in filled polymers: A fractional calculus approach, J. Chem. Phys. 103 (1995) 7180–7186.

16. L. Debnath, Recent applications of fractional calculus to science and engineering, Int. J. Math. Math. Sci. 54 (2003) 3413–3442.

17. A. Oustaloup, V. Pommier, P. Lanusse, Design of a fractional control using performance contours. Application to an electromechanical system, Fract. Calc. Appl. Anal. 6 (1) (2003) 1–24.

18. K. Diethelm, A.D. Freed, On the solution of nonlinear fractional order differential equations used in the modeling of viscoplasticity, in: F. Keil, W. Mackens, H. Voss, J. Werther (Eds.), Scientific Computing in Chemical Engineering II-Computational Fluid Dynamics, Reaction Engineering and Molecular Properties, Springer-Verlag, Heidelberg, 1999, pp. 217–224.

19. R. Hilfer, Applications of Fractional Calculus in Physics, World Scientific, Singapore, 2000.

20. J.A. Tenreiro Machado, R.S. Barbosa, Introduction to the special issue on fractional differentiation and its applications, J. Vib. Control 9–10 (2008) 1253.

21. A.A. Kilbas, H.M. Srivastava, J.J. Trujillo, Theory and Applications of Fractional Differential Equations, Elsevier, Amsterdam, 2006.

22. K.B. Oldham, J. Spanier, The Fractional Calculus, Academic Press A subsidiary of Harcourt Brace Jovanovich, Publishers., New York, 1974.

23. I. Podlubny, Fractional Differential Equations, Academic Press, Inc, San Diego, CA, 1999.

24. S.G. Samko, A.A. Kilbas, O.I. Marichev, Fractional Integrals and Derivatives, Gordon and Breach Science Publishers, Yverdon, 1993.

25. B. Ross, S.G. Samko, E.R. Love, Functions that have no first order derivative might have fractional derivatives of all orders less than one, Real Annal. Exchange 20 (1) (1994/95) 140–157.

CITATION

Ricardo Almeida, Delfim F.M. Torres, Calculus of variations with fractional derivatives and fractional integrals, Applied Mathematics Letters, Volume 22, Issue 12, December 2009, Pages 1816-1820, ISSN 0893-9659, http://dx.doi.org/10.1016/j.aml.2009.07.002.

Extensions of Certain Classical Integrals of Erdélyi for Gauss Hypergeometric Functions

C.M. Joshi[a] and Yashoverdhan Vyas[b]

[a]106, Arihant Nagar, Kalka Mata Road,
Udaipur 313001, Rajasthan, India
[b]Department of Mathematics and Statistics,
College of Science, M. L. Sukhadia
University, Udaipur 313001, Rajasthan, India

ABSTRACT

It is shown how series manipulation technique and certain classical summation theorems for hypergeometric series can be used to prove Erdélyi's integral representations for $_2F_1(z)$, originally proved using fractional calculus. The method not only leads to generalizations but also leads to new integrals of Erdélyi type for certain $_{q+1}Fq(z)$ and corresponding Pochhammer contour integrals. The technique outlined here, compared to the method of fractional calculus, seems to be more effective as it not only provides transparent elementary proofs of Erdélyi's integrals but even leads to various generalizations.

INTRODUCTION

Although the integrals involving and representing hypergeometric functions have numerous applications in pure and applied mathematics (see, for example, [6]), not all such integrals have been collected in tables or are readily available in the mathematical literature. In this context, we shall consider the integrals of Erdélyi type.

Let $z \neq 1$ and $|\arg(1-z)| < \pi$. In 1939, Erdélyi [4] showed that Euler›s integral

$$_2F_1 \begin{bmatrix} \alpha, \beta; \\ \gamma; \end{bmatrix} z = \frac{\Gamma(\gamma)}{\Gamma(\beta)\Gamma(\gamma - \beta)} \int_0^1 t^{\beta-1}(1 - t)^{\gamma-\beta-1}(1 - tz)^{-\alpha}\, dt,$$

(1.1)

where $\mathrm{Re}(\gamma) > R(\beta) > 0$, and Bateman's [1] extension of (1.1)

$$_2F_1 \begin{bmatrix} \alpha, \beta; \\ \gamma; \end{bmatrix} z = \frac{\Gamma(\gamma)}{\Gamma(\mu)\Gamma(\gamma - \mu)} \int_0^1 t^{\mu-1}(1 - t)^{\gamma-\mu-1} {}_2F_1 \begin{bmatrix} \alpha, \beta; \\ \mu; \end{bmatrix} zt\, dt,$$

(1.2)

where $R(\gamma) > \mathrm{Re}(\mu) > 0$, have extensions of the forms

$$_2F_1 \begin{bmatrix} \alpha, \beta; \\ \gamma; \end{bmatrix} z = \frac{\Gamma(\gamma)}{\Gamma(\mu)\Gamma(\gamma - \mu)} \int_0^1 t^{\mu-1}(1 - t)^{\gamma-\mu-1}(1 - tz)^{\lambda-\alpha-\beta}$$

$$\times {}_2F_1 \begin{bmatrix} \lambda - \alpha, \lambda - \beta; \\ \mu; \end{bmatrix} tz \; {}_2F_1 \begin{bmatrix} \alpha + \beta - \lambda, \lambda - \mu; \frac{(1 - t)z}{1 - tz} \\ \gamma - \mu; \end{bmatrix} dt,$$

(1.3)

where $\mathrm{Re}(\gamma) > \mathrm{Re}(\mu) > 0$,

$$_2F_1 \begin{bmatrix} \alpha, \beta; \\ \gamma; \end{bmatrix} z = \frac{\Gamma(\gamma)}{\Gamma(\mu)\Gamma(\gamma - \mu)} \int_0^1 t^{\mu-1}(1 - t)^{\gamma-\mu-1}(1 - tz)^{-\alpha'}$$

$$\times {}_2F_1 \begin{bmatrix} \alpha - \alpha', \beta; \\ \mu; \end{bmatrix} tz \; {}_2F_1 \begin{bmatrix} \alpha', \beta - \mu; \frac{(1 - t)z}{1 - tz} \\ \gamma - \mu; \end{bmatrix} dt,$$

(1.4)

where $\mathrm{Re}(\gamma) > \mathrm{Re}(\mu) > 0$ and

$$_2F_1 \begin{bmatrix} \alpha, \beta; \\ \gamma; \end{bmatrix} z = \frac{\Gamma(\gamma)\Gamma(\mu)}{\Gamma(\lambda)\Gamma(\nu)\Gamma(\gamma + \mu - \lambda - \nu)} \int_0^1 t^{\nu-1}(1 - t)^{\gamma+\mu-\lambda-\nu-1}$$

$$\times {}_2F_1 \begin{bmatrix} \mu - \lambda, \gamma - \lambda; \\ \gamma + \mu - \lambda - \nu; \end{bmatrix} 1 - t \; {}_3F_2 \begin{bmatrix} \alpha, \beta, \mu; \\ \lambda, \nu; \end{bmatrix} tz\, dt,$$

(1.5)

where $\mathrm{Re}(\lambda,\nu,\gamma+\mu-\lambda-\nu)>0$. It may be noted that, when $\mu=\beta$, (1.2) becomes (1.1) and when $\lambda=\alpha+\beta$, $\alpha'=0$ and $\lambda=\mu$, then, and, respectively, get converted into (1.2). Erdélyi also considered special cases and confluent limit cases of his formulas and some formulas obtained by applying transformation formulas to the hypergeometric functions in the integrand.

Gasper [7] pointed out some important applications of Erdélyi›s fractional integral (1.3), such as to derive Dirichlet–Mehler-type integral representations for Jacobi polynomials and for generalized Legendre functions, and to prove the positivity of certain sums of generalized Legendre functions and derived the discrete analogue of (1.3). Later in [8], this work was extended for (1.4) and (1.5) to q-analogues, along with the derivation of a generalization of a q-Kampé de Fériet sum, which was conjectured in the work [14] on the evaluation of the 9−j recoupling coefficients appearing in the quantum theory of angular momentum.

On the other hand, as recorded in [12, pp. 287–288], Erdélyi [3] gave a general form for (1.2) as

$$
{}_pF_q\left[\begin{array}{c} a_1,\ldots,a_p; \\ b_1,\ldots,b_q; \end{array} z \right]
$$

$$
=\prod_{j=1}^{m}\left\{\frac{\Gamma(\alpha_j)}{\Gamma(a_j)\Gamma(\alpha_j-a_j)}\right\}\prod_{j=1}^{n}\left\{\frac{\Gamma(b_j)}{\Gamma(\beta_j)\Gamma(b_j-\beta_j)}\right\}
$$

$$
\times\int_0^1\cdots\int_0^1\prod_{j=1}^{m}\{u_j^{a_j-1}(1-u_j)^{\alpha_j-a_j-1}\}\prod_{j=1}^{n}\{v_j^{b_j-1}(1-v_j)^{b_j-\beta_j-1}\}
$$

$$
\times\,{}_pF_q\left[\begin{array}{c} \alpha_1,\ldots,\alpha_m,a_{m+1},\ldots,a_p; \\ \beta_1,\ldots,\beta_n,b_{n+1},\ldots,b_q; \end{array} zu_1\ldots u_m v_1\ldots v_n \right]\,du_1\ldots du_m\,dv_1\ldots dv_n,
$$

$$(1.6)$$

$m \leq p; n \leq q; Re\left(\alpha_j\right) > Re\left(\alpha_j\right) > 0, j = 1,....,m; Re\left(\beta_j\right) > Re(bj) > 0, j = 1,...,n; p \leq q$

and $|z| < \infty$ (or $p = q+1, z \neq 1$ and $|arg(1-z)| < \pi$), which, for m=0 and n=1, yields the elegant result

$$
{}_pF_q\left[\begin{array}{c} a_1,...,a_p; \\ b_1,...,b_q; \end{array} z\right] = \frac{\Gamma(b_1)}{\Gamma(\lambda)\Gamma(b_1 - \lambda)} \int_0^1 t^{\lambda-1}(1-t)^{b_1-\lambda-1} {}_pF_q\left[\begin{array}{c} a_1,...,a_p; \\ \lambda, b_2,...,b_q; \end{array} zt\right] dt, \quad (1.7)
$$

$p \leq q$ and $|z| < \infty$ $(p = q+1, z \neq 1$ and $|arg(1-z)| < \pi); Re(b_1) > Re(\lambda) > 0$, whose special case, when p=2 and q=1, corresponds to (1.2). This work on (1.2) has been further extended by many authors (see [12, p. 288]) for Appell, Kampé de Fériet, Lauricella and other multiple series by using fractional calculus.

Erdélyi used fractional integration by parts and transformation formulas for ${}_2F_1$ hypergeometric functions to derive his integrals (1.3)–(1.5).

In this paper, the series manipulation technique and the classical summation theorems are used to give an alternative way of proof for Erdélyi integrals (1.3)–(1.5) and Motivated, from the above way of proof, in Section 3, new integrals of Erdélyi type for certain ${}_{q+1}F_q(z)$ are conjectured and proved. A generalization and unification of (1.3), (1.4), and (3.1) is obtained in Section 4(i). Multidimensional cases of integrals (1.3), (1.4), (3.1), (4.1) and of integrals obtained in Section 3, are contained in Section 4(ii). A generalization of (1.5), in terms of a most general power series is found, whose particular case in terms of ${}_pF_q(z)$, includes (1.7) as a particular case, is obtained in Section 4(iii). Furthermore, in Section 4(iv), we are able to give an interesting multidimensional case of the generalization of (1.5), whose particular case in terms of ${}_pF_q(z)$, includes the integral (1.6) in its fold. In Section 5, it is shown that how the method outlined in this paper can be used to prove the corresponding Pochhammer contour analogues of all the extensions of Erdélyi's integrals developed here.

ALTERNATIVE PROOFS FOR THE ERDÉLYI'S INTEGRALS (1.3)–(1.5)

i. Let us denote the right-hand side of (1.3) by I, then replacing the $_2F_1$'s of the integrand by their series form and interchanging the order of summations and integration, which is valid when $|z|<1$, we get

$$
I = \frac{\Gamma(\gamma)}{\Gamma(\mu)\Gamma(\gamma - \mu)} \sum_{m,n=0}^{\infty} \frac{(\lambda - \alpha)_m (\lambda - \beta)_m (\lambda - \mu)_n (\alpha + \beta - \lambda)_n}{(\mu)_m (\gamma - \mu)_n} \frac{z^m}{m!} \frac{z^n}{n!}
$$

$$
\times \int_0^1 t^{\mu+m-1}(1 - t)^{\gamma-\mu+n-1}(1 - tz)^{\lambda-\alpha-\beta-n}\, dt.
$$

(2.1)

By using (1.1), with the prescribed conditions, (2.1) becomes

$$
I = \sum_{m,n=0}^{\infty} \frac{(\lambda - \alpha)_m (\lambda - \beta)_m (\lambda - \mu)_n (\alpha + \beta - \lambda)_n (\mu)_m (\gamma - \mu)_n}{(\mu)_m (\gamma - \mu)_n (\gamma)_{m+n}} \frac{z^m}{m!} \frac{z^n}{n!}
$$

$$
\times\ _2F_1 \left[\begin{array}{c} \alpha + \beta - \lambda + n, \mu + m; \\ \gamma + m + n; \end{array} z \right]
$$

or

$$
I = \sum_{k,m,n=0}^{\infty} \frac{(\alpha + \beta - \lambda)_{k+n} (\mu)_{k+m} (\lambda - \alpha)_m (\lambda - \beta)_m (\lambda - \mu)_n}{(\gamma)_{k+m+n} (\mu)_m} \frac{z^k}{k!} \frac{z^m}{m!} \frac{z^n}{n!}.
$$

(2.2)

Now applying the series manipulation technique [10, p. 56, Lemma 10] or [13, p. 100, (1)] twice or more explicitly, using the following triple series manipulation:

$$
\sum_{k=0}^{\infty}\sum_{m=0}^{\infty}\sum_{n=0}^{\infty} A(k,m,n) = \sum_{k=0}^{\infty}\sum_{m=0}^{k}\sum_{n=0}^{k-m} A(k - m - n, m, n)
$$

on the triple series of (2.2), we can write

$$I = \sum_{k=0}^{\infty} \sum_{m=0}^{k} \frac{(\alpha + \beta - \lambda)_{k-m}(\mu)_k(\lambda - \alpha)_m(\lambda - \beta)_m(-k)_m(-1)^m z^k}{(\gamma)_k(\mu)_m m!} {}_2F_1 \left[\begin{array}{c} \lambda - \mu, m - k; \\ 1 - \mu - k; \end{array} 1 \right].$$

(2.3)

After summing the inner ${}_2F_1$ of (2.3) by the Vandermonde theorem, [11, p. 243, (III. 4)] we can write

$$I = \sum_{k=0}^{\infty} \frac{(\lambda)_k(\alpha + \beta - \lambda)_k}{(\gamma)_k} \frac{z^k}{k!} {}_3F_2 \left[\begin{array}{c} \lambda - \alpha, \lambda - \beta, -k; \\ 1 + \lambda - \alpha - \beta - k, \lambda; \end{array} 1 \right].$$

The above inner ${}_3F_2(1)$ can be summed by the Saalschütz theorem [11, p. 243, (III. 2)] to give

$$I = {}_2F_1 \left[\begin{array}{c} \alpha, \beta; \\ \gamma; \end{array} z \right].$$

Thus the proof of (1.3) is completed.

ii. Similarly, in the right-hand side of (1.4), replacing the ${}_2F_1$'s of the integrand by their series form and using (1.1), we obtain a triple series. Applying the series manipulation technique, on this triple series, twice, we can take the first inner series as a ${}_2F_1$ and sum it by the Vandermonde theorem to have a double series, from which again, we can take the inner series as a ${}_2F_1$ and sum it by the Vandermonde theorem and get the ${}_2F_1$, which on the left of (1.4) and thus, we complete the proof of (1.4).

iii. The proof of (1.5) is slightly different and simple. Let the right-hand side of (1.5) be I. Replacing ${}_2F_1$ and ${}_3F_2$ of the integrand by their series form and using Euler's Beta integral, we can have

$$I = \frac{\Gamma(\gamma)\Gamma(\mu)}{\Gamma(\lambda)\Gamma(\gamma+\mu-\lambda)} \sum_{n=0}^{\infty} \frac{(\alpha)_n(\beta)_n(\mu)_n}{(\gamma+\mu-\lambda)_n(\lambda)_n} \frac{z^n}{n!} {}_2F_1\left[\begin{array}{c} \mu-\lambda,\gamma-\lambda; \\ \gamma+\mu-\lambda+n; \end{array} 1\right]$$

Summing the above inner ${}_2F_1$ by the Gauss theorem [11, p. 243 (III.3)], we get that

$$I = {}_2F_1\left[\begin{array}{c} \alpha,\beta; \\ z \\ \gamma; \end{array}\right]$$

and thus the proof of (1.5) is completed.

INTEGRALS OF ERDÉLYI TYPE FOR CERTAIN q+1FQ(Z)

Motivated from the above way of proof of Erdélyi's integrals which uses the Gauss, Vandermonde and Saalschütz theorems, similar integrals are conjectured below, whose proof uses other summation theorems as well. We call these "Erdélyi-type integrals" because they follow the similar method of proof and are integrals of the combination of powers and hypergeometric functions, like (1.3)–(1.5). For example,

$$_3F_2\left[\begin{array}{c} \alpha,\beta,v+\alpha'; \\ z \\ \gamma,v+\alpha; \end{array}\right] = \frac{\Gamma(\gamma)}{\Gamma(\lambda)\Gamma(\gamma-\lambda)} \int_0^1 t^{\lambda-1}(1-t)^{\gamma-\lambda-1}(1-tz)^{-\alpha'}$$

$$\times {}_3F_2\left[\begin{array}{c} \alpha-\alpha',\beta,v; \\ zt \\ \lambda,v+\alpha; \end{array}\right] {}_2F_1\left[\begin{array}{c} \alpha',\beta-\lambda; \frac{z(1-t)}{1-tz} \\ \gamma-\lambda; \end{array}\right] dt,$$

$$(3.1)$$

where $\left|\arg(1-z)\right| < \pi, \operatorname{Re}(\gamma) > \operatorname{Re}(\gamma) > 0$ and $z\neq1$. As $v\to\infty$ (3.1) becomes (1.4) and can be proved by following the method explained in Section 2(i) and the proof will use the Vandermonde and Saalschütz theorems. The other integrals are

$$= \frac{\Gamma(\gamma)}{\Gamma(\alpha+\beta)\Gamma(\gamma-\alpha-\beta)} \int_0^1 t^{\gamma-\alpha-\beta-1}(1-t)^{\alpha+\beta-1}(1+tz)^{-\beta}$$

$$\times {}_2F_1\left[\begin{array}{c} \alpha-\beta, \gamma-\beta; \\ \gamma-\alpha-\beta; \end{array} zt\right] {}_3F_2\left[\begin{array}{c} \alpha, \dfrac{\beta}{2}, \dfrac{1+\beta}{2}; \\ \dfrac{\alpha+\beta}{2}, \dfrac{1+\alpha+\beta}{2}; \end{array} \dfrac{-z(1-t)^2}{(1+tz)^2}\right] dt,$$

$$(3.2)$$

where $|\arg(1-z)| < \pi$; $\mathrm{Re}(\gamma) > \mathrm{Re}(\alpha+\beta) > 0$, and $z \neq 1$. It can be proved by using the method of Section 2(i) and the proof shall use the Saalschütz theorem twice.

$$ {}_3F_2\left[\begin{array}{c} \alpha, \beta, 1+\gamma-\mu-\lambda; \\ 1+\gamma-\mu, 1+\gamma-\lambda; \end{array} z\right]$$

$$= \frac{\Gamma(1+\gamma)}{\Gamma(\beta)\Gamma(1+\gamma-\beta)} \int_0^1 t^{\beta-1}(1-t)^{\gamma-\beta}(1-tz)^{-\alpha}$$

$$\times {}_5F_4\left[\begin{array}{c} \gamma, 1+\dfrac{\gamma}{2}, \mu, \lambda, \alpha; \\ \dfrac{\gamma}{2}, 1+\gamma-\mu, 1+\gamma-\lambda, 1+\gamma-\beta; \end{array} \dfrac{-t(1-t)}{1-tz}\right] dt,$$

$$(3.3)$$

where, $|\arg(1-z)| < \pi$; $\mathrm{Re}(1+\gamma) > \mathrm{Re}(\beta) > 0, z \neq 1$. For proving (3.3), in the right-hand side of it, replacing the ${}_5F_4$ of the integrand by the series form and using (1.1), we get a double series. Applying series manipulation, once, on this double series and summing the resulting inner series by the Dougall's theorem [11, p. 243, (III.14)] we get a ${}_3F_2$ which is on the left of (3.3) and, thus, we complete the proof of (3.3).

$$
{}_4F_3\left[\begin{array}{c} \alpha,\beta,\dfrac{\gamma-\mu}{2},\dfrac{1+\gamma-\mu}{2}; \\[2mm] \gamma-\mu,\dfrac{\gamma}{2},\dfrac{1+\gamma}{2}; \end{array} z\right]
$$

$$
=\frac{\Gamma(\gamma)}{\Gamma(\beta)\Gamma(\gamma-\beta)}\int_0^1 t^{\beta-1}(1-t)^{\gamma-\beta-1}(1-tz)^{-\alpha}
$$

$$
\times {}_2F_1\left[\begin{array}{c} \alpha,\mu; \\ \gamma-\beta; \end{array} \frac{-t(1-t)z}{1-tz}\right] dt,
$$

(3.4)

where $\left|\arg(1-z)\right|<\pi, \mathrm{Re}(\gamma)>\mathrm{Re}(\beta)>0$ and $z\neq1$.

To prove (3.4), in the right-hand side of it we replace $(1-tz)^{-\alpha}$ by $(1-tz)^{-\mu}(1-tz)^{\mu-\alpha}$ and write the ${}_2F_1$ and $(1-tz)^{\mu-\alpha}$ in series, followed by the use of Euler›s integral (1.1) to obtain a triple series. On which, applying series manipulation twice, followed by the use of the Vandermonde and Saalschütz theorems, we get the left-hand side of (3.4) and we thus, complete the proof of (3.4).

$$
{}_4F_3\left[\begin{array}{c} \alpha,\beta,\gamma,1+\alpha+\mu-\beta-\gamma; \\ 1+\alpha-\beta,1+\alpha-\gamma,\beta+\gamma-\mu; \end{array} z\right]
$$

$$
=\frac{\Gamma(1+\alpha-\mu)}{\Gamma(\lambda)\Gamma(1+\alpha-\lambda-\mu)}\int_0^1 t^{\lambda-1}(1-t)^{\alpha-\mu-\lambda}(1-tz)^{-\mu}{}_2F_1\left[\begin{array}{c} \mu,\alpha-\lambda; \\ 1+\alpha-\mu-\lambda; \end{array} \frac{(1-t)z}{1-tz}\right]
$$

$$
\times {}_7F_6\left[\begin{array}{c} \alpha-\mu,1+\dfrac{\alpha-\mu}{2},\beta-\mu,\gamma-\mu,1+\alpha-\beta-\gamma,\dfrac{\alpha}{2},\dfrac{1+\alpha}{2}; \\[2mm] \dfrac{\alpha-\mu}{2},1+\alpha-\beta,1+\alpha-\gamma,\beta+\gamma-\mu,\dfrac{\lambda}{2},\dfrac{1+\lambda}{2}; \end{array} zt^2\right] dt,
$$

(3.5)

where $\left|\arg(1-z)\right|<\pi, \mathrm{Re}(1+\alpha-\mu)>\mathrm{Re}(\lambda)>0$ and $z\neq1$. Eq. (3.5) can be proved by using the method of Section 2(i) and the proof will use the Vandermonde and Dougall theorems.

$$
{}_5F_4\left[\begin{array}{c} \beta,\gamma,1+\alpha-\mu-\lambda,\dfrac{1+\alpha}{2},\dfrac{2+\alpha}{2}; \\[2mm] 1+\alpha-\mu,1+\alpha-\lambda,\dfrac{\beta+\lambda}{2},\dfrac{1+\beta+\lambda}{2}; \end{array}\; z\right]
$$

$$
=\frac{\Gamma(\beta+\gamma)}{\Gamma(\beta)\Gamma(\gamma)}\int_0^1 t^{\gamma-1}(1-t)^{\beta-1}(1-tz)^{\gamma-\alpha-1}(1-t^2z)^{1+\alpha-\beta-\gamma}
$$

$$
\times {}_4F_3\left[\begin{array}{c} \alpha,1+\dfrac{\alpha}{2},\mu,\lambda; \\[2mm] \dfrac{\alpha}{2},1+\alpha-\mu,1+\alpha-\lambda; \end{array}\; \frac{t(1-t)z}{1-tz}\right]dt,
$$

$$(3.6)$$

where $\left|\arg(1-z)\right|<\pi, \mathrm{Re}(\gamma)>0$ and $z\neq 1$. Eq. (3.6) can be proved by the method explained in the Section 2(i) and the proof will use the Saalchütz and Dougall's ${}_5F_4(1)$ [11, p. 243, (III.13)]theorems.

$$
{}_6F_5\left[\begin{array}{c} \alpha,\beta,\gamma,1+2\alpha-\beta-\gamma-\mu,\dfrac{1+\mu}{2},\dfrac{2+\mu}{2}; \\[2mm] 1+\alpha-\beta,1+\alpha-\gamma,\beta+\gamma+\mu-\alpha,\dfrac{\lambda}{2},\dfrac{1+\lambda}{2}; \end{array}\; z\right]
$$

$$
=\frac{\Gamma(\lambda)}{\Gamma(\alpha)\Gamma(\lambda-\alpha)}\int_0^1 t^{\alpha-1}(1-t)^{\lambda-\alpha-1}(1-tz)^{\mu-\alpha}
$$

$$
\times {}_5F_4\left[\begin{array}{c} \mu,1+\dfrac{\mu}{2},\beta+\mu-\alpha,\gamma+\mu-\alpha,1+\alpha-\beta-\gamma; \\[2mm] \dfrac{\mu}{2},1+\alpha-\beta,1+\alpha-\gamma,\beta+\gamma+\mu-\alpha; \end{array}\; zt^2\right]
$$

$$
\times {}_2F_1\left[\begin{array}{c} \alpha-\mu,\lambda-\mu-1;\\[2mm] \lambda-\alpha; \end{array}\; \frac{-zt(1-t)}{1-tz}\right]dt,
$$

$$(3.7)$$

where $\left|\arg(1-z)\right| < \pi, \mathrm{Re}(\lambda) > \mathrm{Re}(\alpha) > 0$ and $z \neq 1$. Eq. (3.7) can be proved by the method explained in Section 2(i) and the proof will use the Vandermonde and Dougall theorems.

The $\beta=0$ case of (3.2) gives an interesting integral representation for binomial function $_1F_0(z)$, involving $_2F_0(z)$. Integrals (3.3) and (3.4) provide new extensions of well-known Euler's integral (1.1), which follows from (3.3), when $\mu=0$ or $\lambda=0$ and from (3.4), when $\mu=0$.

It will be of interest to mention here that following the argument of Rainville [10, pp. 48–49], we can put z=1 in the above mentioned integrals provided the convergence condition of the series $q+1F_q(z)$, on the LHS, for $|z|=1$, is being satisfied. Taking z=1 in (3.3) and using Euler's Beta integral, we get a well-known transformation of a well-poised $_6F_5(-1)$ into a $3F_2(1)$. Taking z=1 in (3.4) and using Euler's Beta integral, we get a known transformation of a $_4F_3(1)$ into a $_3F_2(-1)$ due to Whipple. Taking z=1 in (3.5) and using the Gauss theorem and Euler's Beta integral, we get a new transformation of a well-poised $_4F_3(1)$ into a well-poised $_7F_6(1)$. Taking z=1, in (3.6) and using Euler's Beta integral, we get two new transformations of special double Srivastava–Daoust series [12] into $_5F_4(1)$ and $_5F_6(1)$, respectively. Hence, the integrals of this section generalize certain hypergeometric transformations.

GENERALIZATIONS OF ERDÉLYI'S INTEGRALS

i. Here, we give an interesting unification and generalization of both the Erdélyi's integrals (1.3) and (1.4) along with (3.1).

The integral established is

$$_3F_2\left[\begin{array}{c} v, \xi, \lambda; \\ \gamma, \delta; \end{array} z\right] = \frac{\Gamma(\gamma)}{\Gamma(\mu)\Gamma(\gamma-\mu)} \int_0^1 t^{\mu-1}(1-t)^{\gamma-\mu-1}(1-tz)^{\delta-v-\xi}$$

$$\times \,_3F_2\left[\begin{array}{c} \delta-\xi, \delta-v, \lambda; \\ \mu, \delta; \end{array} zt\right] \,_2F_1\left[\begin{array}{c} \lambda-\mu, v+\xi-\delta; \ (1-t)z \\ \gamma-\mu; \end{array} \frac{(1-t)z}{1-tz}\right] dt,$$

$$(4.1)$$

where $\left|\arg(1-z)\right| < \pi, \mathrm{Re}(\lambda) > \mathrm{Re}(\mu)$ and $z \neq 1$. The proof follows on the lines of the method explained in Section 2(i) and the proof will use the Vandermonde and Saalchütz theorems. It may be noted that in (4.1), when $v=\alpha, \xi=\beta, \delta=\lambda$, integral (1.3) follows; when $\xi=\delta+\alpha'-\alpha, v=\alpha, \lambda=\beta$ and then as $\delta \to \infty$ integral (1.4) follows and when $\xi=\alpha, \lambda=\beta, v=v+\alpha', \delta=v+\alpha$, integral (3.1) follows. Again, using the argument of Rainville [10, pp. 48–49], we can put z=1 in this integral. Taking z=1 and using the Gauss theorem and Euler›s Beta integral, (4.1) converts into the well-known Kummer–Thomae–Whipple $_3F_2(1)$transformation. Thus (4.1) is a generalization of Kummer–Thomae–Whipple $_3F_2(1)$ transformation.

ii. A multidimensional case of Euler's integral (1.1) follows in the form

$$_{q+1}F_q\left[\begin{matrix} a_1,\ldots,a_{q+1}; \\ b_1,\ldots,b_q; \end{matrix} z\right] = \prod_{j=1}^{q}\left\{\frac{\Gamma(b_j)}{\Gamma(a_j)\Gamma(b_j-a_j)}\right\}\int_0^1\cdots\int_0^1\prod_{j=1}^{q}\{t_j^{a_j-1}(1-t_j)^{b_j-a_j-1}\}$$

$$\times(1-t_1\ldots t_q z)^{-a_{q+1}}\,dt_1\ldots dt_q, \tag{4.2}$$

where $\left|\arg(1-z)\right| < \pi, \mathrm{Re}\left(b_j\right) > \mathrm{Re}\left(\alpha_j\right) > 0, j = 1,\ldots.q..$

From the repeated application of the functional equation [10, p. 85, Theorem 28]

$$_pF_q\left[\begin{matrix} a_1,\ldots,a_p; \\ b_1,\ldots,b_q; \end{matrix} z\right] = \frac{\Gamma(b_1)}{\Gamma(a_1)\Gamma(b_1-a_1)}\int_0^1 t^{a_1-1}(1-t)^{b_1-a_1-1}\,_{p-1}F_{q-1}\left[\begin{matrix} a_2,\ldots,a_p; \\ b_2,\ldots,b_q; \end{matrix} zt\right]dt \tag{4.3}$$

and then using the binomial theorem.

The multidimensional case of (4.1) admits the form

$$
{}_{q+2}F_{q+1}\left[\begin{array}{c} a_1,\ldots,a_{q-1},v,\xi,\lambda; \\ b_1,\ldots,b_{q-1},\gamma,\delta; \end{array} z\right]
$$

$$
=\prod_{j=1}^{q-1}\left\{\frac{\Gamma(b_j)}{\Gamma(a_j)\Gamma(b_j-a_j)}\right\}\frac{\Gamma(\gamma)}{\Gamma(\mu)\Gamma(\gamma-\mu)}
$$

$$
\times \int_0^1\cdots\int_0^1\prod_{j=1}^{q-1}\{(t_j)^{a_j-1}(1-t_j)^{b_j-a_j-1}\}
$$

$$
\times t_q^{\mu-1}(1-t_q)^{\gamma-\mu-1}(1-t_1\ldots t_q z)^{\delta-\xi-v}
$$

$$
\times {}_3F_2\left[\begin{array}{c} \delta-\xi,\delta-v,\lambda; \\ \mu,\delta; \end{array} t_1\ldots t_q z\right]
$$

$$
\times {}_2F_1\left[\begin{array}{c} \lambda-\mu,v+\xi-\delta; \\ \gamma-\mu; \end{array} \frac{t_1\ldots t_{q-1}(1-t_q)z}{(1-t_1\ldots t_q z)}\right] dt_1\ldots dt_q,
$$

$$
(4.4)
$$

where $\left|\arg(1-z)\right|<\pi,\operatorname{Re}(\gamma)>\operatorname{Re}(\mu)>0,z\neq 1$ and $\operatorname{Re}(b_j)>\operatorname{Re}(\alpha_j)>0.j=1,\ldots..q=1$. For q=1, (4.4) becomes (4.1). The proof of (4.4) follows along the lines of the method, explained in Section 2(i), but uses (4.2) in place of (1.1) and the used summation theorems will be of the Vandermonde and Saalschütz. Obviously, the multidimensional cases of Erdélyi's integrals and will follow from (4.4), since (4.1) is a generalization and unification of(1.3) and (1.4). Similarly, multidimensional cases of the integrals discussed in Section 3 can also be developed.

iii. Motivated from the alternative proof of Erdélyi's integral (1.5), given in Section 2(iii), it is possible to suggest a general setting for the integral in the form

$$\sum_{n=0}^{\infty} c_n \frac{(\lambda)_n(v)_n}{(\gamma)_n(\mu)_n} z^n = \frac{\Gamma(\gamma)\Gamma(\mu)}{\Gamma(\lambda)\Gamma(v)\Gamma(\gamma + \mu - \lambda - v)}$$

$$\times \int_0^1 t^{v-1}(1-t)^{\gamma+\mu-\lambda-v-1} {}_2F_1 \begin{bmatrix} \mu - \lambda, \gamma - \lambda; \\ \gamma + \mu - \lambda - v; \end{bmatrix} 1 - t, \end{bmatrix} \phi(zt)\,dt,$$

(4.5)

where $\mathrm{Re}(\lambda, v, \gamma + \mu - \lambda - v) > 0$ and $\varphi(z)$ is assumed to be a convergent series defined by

$$\phi(z) = \sum_{n=0}^{\infty} c_n z^n,$$

(4.6)

where c_n is any bounded sequence of complex numbers. Following the method of proof for (1.5) given inSection 2(iii), (4.5) can also be proved.

If we take

$$c_n = \frac{(\alpha)_n(\beta)_n(\mu)_n}{(\gamma)_n(v)_n n!},$$

then (4.5) leads us to (1.5).

If we choose $\gamma = b_1$ and

$$c_n = \frac{(a_1)_n \dots (a_p)_n(\mu)_n}{(b_2)_n \dots (b_q)_n(\lambda)_n(v)_n n!},$$

then (4.5) gives us

$${}_pF_q \begin{bmatrix} a_1, \dots, a_p; \\ b_1, \dots, b_q; \end{bmatrix} z \end{bmatrix} = \frac{\Gamma(b_1)\Gamma(\mu)}{\Gamma(\lambda)\Gamma(v)\Gamma(b_1 + \mu - \lambda - v)} \int_0^1 t^{v-1}(1-t)^{b_1+\mu-\lambda-v-1}$$

$$\times {}_2F_1 \begin{bmatrix} \mu - \lambda, b_1 - \lambda; \\ b_1 + \mu - \lambda - v; \end{bmatrix} 1 - t \end{bmatrix} {}_{p+1}F_{q+1} \begin{bmatrix} a_1, \dots, a_p, \mu; \\ b_2, \dots, b_q, \lambda, v; \end{bmatrix} zt \end{bmatrix} dt,$$

(4.7)

where $p \leq q$ and $|z| < \infty$ (or $p = q+1$, $z \neq 1$, and $|\arg(1-z)| < \pi$); $Re(\lambda, v, b_1 + \mu - \lambda - v) > 0$. Eq. (4.7) generalizes (1.7) since $\mu = v$ case of (4.7) is (1.7).

iv. At this stage, one can observe that the method of proof used for (4.5) above can be applied any number of times to prove the following multidimensional case of the integral given by (4.5):

$$
\sum_{k=0}^{\infty} c_k \prod_{i=1}^{m} \left\{ \frac{(a_i)_k (\lambda_i)_k}{(\alpha_i)_k (\mu_i)_k} \right\} \prod_{j=1}^{n} \left\{ \frac{(\beta_j)_k (\xi_j)_k}{(b_j)_k (v_j)_k} \right\} z^k
$$

$$
= \prod_{i=1}^{m} \left\{ \frac{\Gamma(\alpha_i)\Gamma(\mu_i)}{\Gamma(a_i)\Gamma(\lambda_i)\Gamma(\alpha_i + \mu_i - a_i - \lambda_i)} \right\}
$$

$$
\times \prod_{j=1}^{n} \left\{ \frac{\Gamma(b_j)\Gamma(v_j)}{\Gamma(\xi_j)\Gamma(\beta_j)\Gamma(b_j + v_j - \xi_j - \beta_j)} \right\}
$$

$$
\times \int_0^1 \cdots \int_0^1 \prod_{i=1}^{m} \left\{ u_i^{\lambda_i - 1}(1-u_i)^{\alpha_i + \mu_i - a_i - \lambda_i - 1} {}_2F_1 \left[\begin{array}{c} \mu_i - a_i, \alpha_i - a_i; \\ \alpha_i + \mu_i - a_i - \lambda_i; \end{array} (1 - u_i) \right] \right\}
$$

$$
\times \prod_{j=1}^{n} \left\{ v_j^{\beta_j - 1}(1-v_j)^{b_j + v_j - \xi_j - \beta_j - 1} {}_2F_1 \left[\begin{array}{c} v_j - \xi_j, b_j - \xi_j; \\ b_j + v_j - \xi_j - \beta_j; \end{array} (1 - v_i) \right] \right\}
$$

$$
\times \phi(z u_1 \ldots u_m \, v_1 \ldots v_n) \, du_1 \ldots du_m \, dv_1 \ldots dv_n,
$$

(4.8)

where

$$
\phi(z) = \sum_{k=0}^{\infty} c_k z^k
$$

is assumed to be a convergent series and c_k is any bounded sequence of complex numbers and

$$
Re(a_i, \lambda_i, \alpha_i + \mu_i - a_i - \lambda_i, \xi_j, \beta_j, b_j + v_j - \xi_j - \beta_j) > 0, \quad i = 1, \ldots, m, \quad j = 1, \ldots, n.
$$

Now in (4.8), setting

$$
c_k = \frac{\prod_{i=1}^{m}\{(\alpha_i)_k (\mu_i)_k\} \prod_{r=1}^{p}(a_{m+r})_k \prod_{j=1}^{n}(v_j)_k}{\prod_{j=1}^{n}\{(\beta_j)_k (\xi_j)_k\} \prod_{l=1}^{q}(b_{n+l})_k \prod_{i=1}^{m}(\lambda_i)_k k!}, \quad m \leq p, \ n \leq q,
$$

we get

$$
{}_pF_q\left[\begin{matrix} a_1,\ldots,a_p; \\ b_1,\ldots,b_q; \end{matrix} z\right]
$$

$$
-\prod_{i=1}^{m}\left\{\frac{\Gamma(\alpha_i)\Gamma(\mu_i)}{\Gamma(a_i)\Gamma(\lambda_i)\Gamma(\alpha_i+\mu_i-a_i-\lambda_i)}\right\}\cdot\prod_{j=1}^{n}\left\{\frac{\Gamma(b_j)\Gamma(v_j)}{\Gamma(\xi_j)\Gamma(\beta_j)\Gamma(b_j+v_j-\xi_j-\beta_j)}\right\}
$$

$$
\cdot\int_0^1\cdots\int_0^1\prod_{i=1}^{m}\left\{u_i^{\lambda_i-1}(1-u_i)^{\alpha_i+\mu_i-a_i-\lambda_i-1}\cdot{}_2F_1\left[\begin{matrix}\mu_i-a_i,\alpha_i-a_i; \\ \alpha_i+\mu_i-a_i-\lambda_i;\end{matrix}(1-u_i)\right]\right\}
$$

$$
\prod_{j=1}^{n}\left\{v_j^{\beta_j-1}(1-v_j)^{b_j+v_j-\xi_j-\beta_j-1}\cdot{}_2F_1\left[\begin{matrix}v_j-\xi_j,b_j-\xi_j; \\ b_j+v_j-\xi_j-\beta_j;\end{matrix}(1-v_j)\right]\right\}{}_{p+m+n}F_{q+m+n}
$$

$$
\cdot\left[\begin{matrix}\alpha_1,\ldots,\alpha_m,a_{m+1},\ldots,a_p,\mu_1,\ldots\mu_m,v_1,\ldots v_n; \\ \beta_1,\ldots,\beta_n,b_{n+1},\ldots,b_a,\lambda_1,\ldots\lambda_m,\xi_1,\ldots\xi_n;\end{matrix} zu_1\ldots u_m v_1\ldots v_n\right] du_1\ldots du_m\, dv_1\ldots dv_n,\qquad(4.9)
$$

where,

$$m\le p; n\le q; \mathrm{Re}(\alpha_i,\lambda_i,\alpha_i+\mu_i-\alpha_i-\lambda_i)>0, i=1,\ldots,m;\ \mathrm{Re}(\xi_j,\beta_j,b_j+v_j-\xi_j-\beta_j)>0; j=1,\ldots,n; p\le q \text{ and } |z|<\infty (\text{or } p=q+1, z\ne 1$$

and $|z|<\infty$ (or p=q+1, z≠1, and $|\arg(1-z)|<\pi$), which for $\mu_i=\lambda_i$ and $v_j=\xi_j, i=1,\ldots\ldots m, j=1,\ldots\ldots,n$ gives the Erdélyi's result (1.6). It may be noted that as (1.6) is a multidimensional extension of (1.2), similarly (4.9) provides a multidimensional case for Erdélyi's extension (1.5).

POCHHAMMER CONTOUR ANALOGUES

i. The convergence conditions for integral (1.1) have been relaxed by taking the integral round Pochhammer's double-loop contour and the result that follows is [5, p. 115(4)]; [11, p. 19, (1.6.18)]; [9, p. 58, (8)]

$$
{}_2F_1\left[\begin{array}{c} \alpha, \beta; \\ \gamma; \end{array} z\right] = \frac{-\Gamma(\gamma)e^{-i\pi\gamma}}{4\Gamma(\beta)\Gamma(\gamma-\beta)\sin(\pi\beta)\sin\{\pi(\gamma-\beta)\}}
$$

$$
\times \int^{(1+,0+,1-,0-)} t^{\gamma-1}(1-t)^{\gamma-\beta-1}(1-tz)^{-\alpha}\,dt
$$

(5.1)

for all arg z and for β and γ−b≠1,2,3,... only.

In this context the method of proof outlined in the previous sections, suggest the corresponding Pochhammer contour analogues of the Erdélyi-type integrals, generalized Erdélyi's integrals and their multidimensional extensions.

For instance, the Pochhammer contour analogue of (3.7) assumes the form

$$
{}_6F_5\left[\begin{array}{c} \alpha, \beta, \gamma, 1+2\alpha-\beta-\gamma-\mu, \dfrac{1+\mu}{2}, \dfrac{2+\mu}{2}; \\ 1+\alpha-\beta, 1+\alpha-\gamma, \beta+\gamma+\mu-\alpha, \dfrac{\lambda}{2}, \dfrac{1+\lambda}{2}; \end{array} z\right]
$$

$$
= \frac{-\Gamma(\gamma)e^{-i\pi\lambda}}{4\Gamma(\alpha)\Gamma(\lambda-\alpha)\sin\pi\alpha\sin\pi(\lambda-\alpha)} \int^{(1+,0+,1-,0-)} t^{\alpha-1}(1-t)^{\lambda-\alpha-1}(1-tz)^{\mu-\alpha}
$$

$$
\times {}_5F_4\left[\begin{array}{c} \mu, 1+\dfrac{\mu}{2}, \beta+\mu-\alpha, \gamma+\mu-\alpha, 1+\alpha-\beta-\gamma; \\ \dfrac{\mu}{2}, 1+\alpha-\beta, 1+\alpha-\lambda, \beta+\gamma+\mu-\alpha; \end{array} zt^2\right]
$$

$$
\times {}_2F_1\left[\begin{array}{c} \alpha-\mu, \lambda-\mu-1; \\ \lambda-\alpha; \end{array} \dfrac{-zt(1-t)}{1-tz}\right]\,dt
$$

(5.2)

for all arg z and for α and λ−α≠1,2,3,... only.

This can be proved by the method explained in Section 2(i) but the proof will use (5.1) instead of (1.1) and the used summation theorems will be of the Vandermonde and Dougall. Similarly, Pochhammer contour analogues of other Erdélyi-type integrals and (4.1) can also be developed.

ii.   The Pochhammer contour analogue of Beta integral [9, p. 18]; [2, p. 214], is given by

$$\int^{(1+,0+,1-,0-)} t^{\alpha-1}(1-t)^{\beta-1}\,dt = \frac{-4\sin(\pi\alpha)\sin(\pi\beta)}{e^{-i\pi(\alpha+\beta)}}\frac{\Gamma(\alpha)\Gamma(\beta)}{\Gamma(\alpha+\beta)},$$

(5.3)

where $\alpha, b \neq 0,1,2,\dots$ only.

As pointed out in [9, p. 61], the contour analogue of (4.3) can be obtained, by using (5.3). We can use the contour analogue of (4.3) to obtain the contour analogue of (4.2), which in turn can be used to develop the contour analogues of (4.4) and the multidimensional cases of Erdélyi-type integrals, on the line of Section 4(ii).

iii.   The contour analogue of (1.5) admits the form

$${}_2F_1\left[\begin{matrix} \alpha, \beta; \\ \gamma; \end{matrix}\,z\right] = \frac{-\Gamma(\gamma)\Gamma(\mu)e^{-i\pi(\gamma+\mu-\lambda)}}{4\sin(\pi v)\sin\{\pi(\gamma+\mu-\lambda-v)\}\Gamma(\lambda)\Gamma(v)\Gamma(\gamma+\mu-\lambda-v)}$$

$$\times \int^{(1+,0+,1-,0-)} t^{v-1}(1-t)^{\gamma+\mu-\lambda-v-1}{}_2F_1\left[\begin{matrix}\mu-\lambda, \gamma-\lambda; \\ \gamma+\mu-\lambda-v;\end{matrix}\,1-t\right]$$

$$\times {}_3F_2\left[\begin{matrix}\alpha, \beta, \mu; \\ \lambda, v;\end{matrix}\,tz\right]\,dt$$

(5.4)

for all $\arg z$ and $\mathrm{Re}(\lambda)>0; v, \gamma-\lambda+\mu-v \neq 1,2\dots$, only.

The proof of (5.4) is on the lines of the method explained in Section 2(iii) but that will use (5.3) in place of beta integral.

Similarly, the generalization and multidimensional extensions of (5.4) can be developed on the lines ofSections 4(iii) and (iv).

CONCLUSIONS

In conclusion, this paper illustrate the superiority of series manipulation technique over the fractional calculus technique, particularly, in the derivations of the Erdélyi-type integrals and their various generalizations. The complete list of such integrals and details on their formation (i.e., the determination of the parameters and the combination of z and t in the integrals) are being prepared and will be communicated soon, in a separate paper. The multiple series analogues and q-series analogues of the results of this paper are also under preparation and will be announced soon.

ACKNOWLEDGMENTS

The second author is thankful to CSIR, New Delhi, India for providing the fellowship.

REFERENCES

1. H.Bateman, The solution of linear di8erential equations by means of deMnite integrals,Trans.Cambridge Philos.Soc.21 (1909) 171–196.

2. E.T. Copson, Theory of Functions of Complex Variable, Oxford University Press, London, New York, 1935.

3. A.Erd &eyli, Der Zusammenhang zwischen verschiedenen Integraldarstellungen hypergeometrischer Funkionen, Quart.J.Math.Oxford 8 (1937) 200–213.

4. A.Erd &eyli, Transformation of hypergeometric integrals by means of fractional integration by parts, Quart.J.Math.Oxford 10 (1939) 176–189.

5. A.Erd &elyi, et al., Higher Transcendental Functions, Vol. I, McGraw-Hill, New York, Toronto, London, 1953.138 C.M. Joshi, Y. Vyas / Journal of Computational and Applied Mathematics 160 (2003) 125–138

6. H.Exton, Handbook of Hypergeometric Integrals, Ellis Horwood Limited, Halsted Press, Wiley, New York, Brisbane,Chichester, Toronto, 1978.

7. G.Gasper, Formulas of Dirichlet–Mehler type, in: Fractional Calculus and its Applications, Proceedings of theInternational Conference, University of New Haven, West Haven Conn., 1974, Lecture Notes in Mathematics, Vol.457, Springer, Berlin, New York, 1975, pp.207–215.

8. G.Gasper, q-Extensions of Erd&elyi's fractional integral representations for hypergeometric functions and somemsummation formulas for double q-Kampe de Feriet series, contemporary mathematics, Amer.Math.Soc.254 (2000)187–198.

9. Y.L.Luke, The Special Functions and their Approximations, Vol.I, Academic Press, New York, London, 1969.

10. E.D. Rainville, Special Functions, Macmillan, New York, 1960.

11. L.J. Slater, Generalized Hypergeometric Functions, Cambridge University Press, Cambridge, London, New York,1966.

12. H.M. Srivastava, P.W. Karlsson, Multiple Gaussian Hypergeometric Series, Ellis Horwood Limited, Halsted Press,Wiley, New York, Brisbane, Chichester, Toronto, 1985.

13. H.M. Srivastava, H.L. Manocha, A Treatise on Generating Functions, Ellis Horwood Limited, Halsted Press, Wiley,New York, Brisbane, Chichester, Toronto, 1984.

14. J.Van der Jeugt, S.N.Pitre, K.Srinivasa Rau, Multiple hypergeometric functions and the 9-j coeKcients, J.Phys.A 27 (1994) 5251–5264.

CITATION

C.M. Joshi, Yashoverdhan Vyas, Extensions of certain classical integrals of Erdélyi for Gauss hypergeometric functions, Journal of Computational and Applied Mathematics, Volume 160, Issues 1–2, 1 November 2003, Pages 125-138, ISSN 0377-0427, http://dx.doi.org/10.1016/S0377-0427(03)00619-8.

A Modified Bessel-type Integral Transform and Its Compositions with Fractional Calculus Operators on Spaces $F_{p,\mu}$ And $F'_{p,\mu}$

H.-J. Glaeske[a], Anatoly A. Kilbas[b], and Megumi Saigo[c]

[a]Department of Mathematics and Informatics, Friedrich Schiller University, D-07740 Jena, Germany
[b]Department of Mathematics and Mechanics, Belarusian State University, Minsk 220050, Belarus
[c]Department of Applied Mathematics, Fukuoka University, Fukuoka 814-0180, Japan

3

ABSTRACT

The paper is devoted to study the integral transform

$$(L_{\gamma,\sigma}^{(\beta)} f)(x) = \int_0^\infty \lambda_{\gamma,\sigma}^{(\beta)}(xt) f(t)\, dt \quad (x > 0)$$

with the kernel

$$\lambda_{\gamma,\sigma}^{(\beta)}(z) = \frac{\beta}{\Gamma(\gamma + 1 - 1/\beta)} \int_1^\infty (t^\beta - 1)^{\gamma - 1/\beta} t^\sigma e^{-zt}\, dt$$

For $\beta > 0$; Re $(\gamma) > 1/\beta - 1$; $\sigma \in \mathbb{R}$; Re$(z) > 0$, which is a generalization of the modified Bessel function of the third kind or Macdonald function $K_\gamma(z)$. Properties of $\lambda_{\gamma,\sigma}^{(\beta)}(z)$ are investigated and compositions of the operator $L_{\gamma,\sigma}^{(\beta)}$ with the left- and right-sided Liouville fractional integrals and derivatives are proved.

INTRODUCTION

The paper deals with the integral transform

$$(L_{\gamma,\sigma}^{(\beta)} f)(x) = \int_0^\infty \lambda_{\gamma,\sigma}^{(\beta)}(xt) f(t) \, dt \quad (x > 0) \tag{1.1}$$

involving the function

$$\lambda_{\gamma,\sigma}^{(\beta)}(z) = \frac{\beta}{\Gamma(\gamma + 1 - 1/\beta)} \int_1^\infty (t^\beta - 1)^{\gamma - 1/\beta} t^\sigma e^{-zt} \, dt$$

$$\left(\beta > 0; \ \mathrm{Re}(\gamma) > \frac{1}{\beta} - 1; \ \sigma \in \mathbb{R}; \ \mathrm{Re}(z) > 0 \right) \tag{1.2}$$

as a kernel. when β=2 and σ=0, then

$$\lambda_{\gamma,0}^{(2)}(z) = \frac{2}{\sqrt{\pi}} \left(\frac{2}{z} \right)^\gamma K_{-\gamma}(z) \quad (\mathrm{Re}(\gamma) > -\tfrac{1}{2}), \tag{1.3}$$

where $K_{-\gamma}(z)$ is the modified Bessel function of the third kind or Macdonald function [3, Section 7.2.2].

Transform (1.1) is a modification of the Bessel-type integral transform

$$(L_\gamma^{(n)} f)(x) = \int_0^\infty \lambda_\gamma^{(n)}(xt) f(t) \, dt \quad (x > 0) \tag{1.4}$$

with the kernel

$$\lambda_\gamma^{(n)}(z) = \frac{(2\pi)^{(n-1)/2} \sqrt{n}}{\Gamma(\gamma + 1 - 1/n)} \left(\frac{z}{n} \right)^{\gamma n} \int_1^\infty (t^n - 1)^{\gamma - 1/n} e^{-zt} \, dt \quad \left(n \in \mathbb{N}; \ \mathrm{Re}(\gamma) > \frac{1}{n} - 1 \right) \tag{1.5}$$

according to the relation

$$(L_\gamma^{(n)} f)(x) = (2\pi)^{(n-1)/2} n^{-(n\gamma + 1/2)} x^{\gamma n} (L_{\gamma,0}^{(n)} t^n f)(x) \tag{1.6}$$

following from (1.1)-(1.2) and (1.4)-(1.5). Transform (1.4), introduced by Krätzel [15], is reduced to the Laplace and Meijer transforms when n=1 and 2, respectively. The properties of $L_\gamma^{(n)}$ such as inversion and convolution theorems, operational rules, differentiation relations, and connections with differential operators were investigated in [15], [16], [17] and [18].

It should be noted that transform (1.4), being useful for the usual differentiation, is not suitable for the fractional one. For this reason transform (1.1) is more preferable. Our paper is devoted to study the compositions of $L_{\gamma,\sigma}^{(\beta)}$ with the left-sided I_{0+}^α, D_{0+}^α and right-sided I_-^α, D_-^α Liouville fractional integrals and derivatives, defined for $\alpha,x>0$ by (see [32, Section 5.1]):

$$(I_{0+}^\alpha \varphi)(x) = \frac{1}{\Gamma(\alpha)} \int_0^x \frac{\varphi(t)\,dt}{(x-t)^{1-\alpha}}, \tag{1.7}$$

$$(D_{0+}^\alpha \varphi)(x) = \left(\frac{d}{dx}\right)^{[\alpha]+1} (I_{0+}^{1-\{\alpha\}} \varphi)(x), \tag{1.8}$$

$$(I_-^\alpha \varphi)(x) = \frac{1}{\Gamma(\alpha)} \int_x^\infty \frac{\varphi(t)\,dt}{(t-x)^{1-\alpha}}, \tag{1.9}$$

$$(D_-^\alpha \varphi)(x) = \left(-\frac{d}{dx}\right)^{[\alpha]+1} (I_-^{1-\{\alpha\}} \varphi)(x), \tag{1.10}$$

where $[\alpha]$ and $\{\alpha\}$ are integral and fractional parts of α, respectively. We prove the relations

$$L_{\gamma,\sigma}^{(\beta)} I_{0+}^\alpha \varphi = x^{-\alpha} L_{\gamma,\sigma-\alpha}^{(\beta)} \varphi, \tag{1.11}$$

$$L_{\gamma,\sigma}^{(\beta)} D_{0+}^\alpha \varphi = x^\alpha L_{\gamma,\sigma+\alpha}^{(\beta)} \varphi, \tag{1.12}$$

$$I_-^\alpha L_{\gamma,\sigma}^{(\beta)} \varphi = L_{\gamma,\sigma-\alpha}^{(\beta)} x^{-\alpha} \varphi, \tag{1.13}$$

$$D_-^\alpha L_{\gamma,\sigma}^{(\beta)} \varphi = L_{\gamma,\sigma+\alpha}^{(\beta)} x^\alpha \varphi \tag{1.14}$$

between the modified Bessel-type integral transform (1.1) and the fractional calculus operators (1.7)-(1.10) in the spaces $F_{p,\mu}$ and $F'_{p,\mu}$ ($1 \le p \le \infty; \mu \in \mathbb{C}$) of test functions and generalized functions developed by McBride [20], [21] and [22] (see also [2]).

For $\mu \in \mathbb{C}$ and $1 \le p \le \infty$ the space $F_{p,\mu}$ is defined by

$$F_{p,\mu} = \left\{ \varphi \in C_0^\infty(\mathbb{R}^+): x^k \frac{d^k}{dx^k}(x^{-\mu}\varphi(x)) \in L^p(\mathbb{R}^+) \ (k \in \mathbb{N}_0) \right\} \tag{1.15}$$

when $1 \le p \le \infty$, and

$$F_{\infty,\mu} = \left\{ \varphi \in C_0^\infty(\mathbb{R}^+): x^k \frac{d^k}{dx^k}(x^{-\mu}\varphi(x)) \to 0 \text{ as } x \to 0 \text{ and } x \to \infty \ (k \in \mathbb{N}_0) \right\}, \tag{1.16}$$

When p=∞, where $\mathbb{N}_0 = \mathbb{N} \cup \{0\} = \{0,1,2,....\}$. $F_{p,\mu}$ is a complete countable multinormed space (Fréchet space) equipped with the topology generated by the family of seminorms in $F_{p,\mu}$ given by

$$\gamma_k^{p,\mu}(\varphi) = \left\| x^k \frac{d^k}{dx^k}(x^{-\mu}\varphi) \right\|_p \qquad (k \in \mathbb{N}_0) \tag{1.17}$$

(see [21, Corollary 2.8]) where $\|\cdot\|_p$ is the usual L_p-norm. The space $F_{p,\mu}$ is closely connected with the Banach space $\mathfrak{L}_{v,r}(v \in \mathbb{C}, 1 \le r < \infty)$ of Lebesgue measurable functions $\varphi(x)$ such that

$$\|\varphi\|_{v,r} = \left(\int_0^\infty |x^v \varphi(x)|^r \frac{dx}{x} \right)^{1/r} < \infty \tag{1.18}$$

developed by Rooney [27] and [28]. $F'_{p,\mu}$ is the space of continuous linear functionals on $F_{p,\mu}$ equipped with the weak topology.

We note that the modified Bessel-type integral transform $L_{\gamma,\sigma}^{(\beta)}$ in (1.1) belongs to the so-called H-transform

$$(Hf)(x) = \int_0^\infty H_{p,q}^{m,n}\left[xt\,\middle|\,\begin{matrix}(a_i,\alpha_i)_{1,p}\\(b_j,\beta_j)_{1,q}\end{matrix}\right] f(t)\,dt \quad (x > 0) \tag{1.19}$$

with the H-function $H_{p,q}^{m,n}\left[z\,\middle|\,\begin{matrix}(a_i,\alpha_i)_{1,p}\\(b_j,\beta_j)_{1,q}\end{matrix}\right]$ as a kernel (see [24, Section 8.1] and [34, Chapter 2]). The Mellin transform defined by

$$(\mathfrak{M}f)(s) = \int_0^\infty f(t)t^{s-1}\,dt \quad (s \in \mathbb{C}) \tag{1.20}$$

has the property

$$\left(\mathfrak{M}\int_0^\infty f(xt)g(t)\,dt\right)(s) = (\mathfrak{M}f)(s)(\mathfrak{M}g)(1-s). \tag{1.21}$$

This relation implies that

$$(\mathfrak{M}Hf)(s) = \frac{\prod_{j=1}^m \Gamma(b_j + \beta_j s)\prod_{i=1}^n \Gamma(1 - a_i - \alpha_i s)}{\prod_{i=n+1}^p \Gamma(a_i + \alpha_i s)\prod_{j=m+1}^q \Gamma(1 - b_j - \beta_j s)}(\mathfrak{M}f)(1-s) \tag{1.22}$$

for integers m, n, p, q such that $0 \le m \le q \le, 0 \le n \le p$ and $a_i b_j \in \mathbb{C}, \alpha_i \beta_j \in \mathbb{R}^+$ $(1 \le i \le p, 1 \le j \le q)$, where an empty product, if it occurs, is taken to be one.

Transform (1.19) includes most of known integral transforms. For the space $\mathfrak{L}_{\nu,r}$ the mapping properties such as the boundedness, the representation and the range of the H-transforms were proved simultaneously in [4], [9], [10], [11] and [1] while the invertibility of (1.19) in $\mathcal{L}_{v,r}$ was given in [33]. By using the results in [4], [9], [10] and [11], the $\mathcal{L}_{v,r}$-theory of the integral transform $L_\gamma^{(n)}$ in (1.4) was constructed in [5]. Mapping properties of the modified H-transforms, generalizing the fractional integration operators, were investigated in [25] and [8], and in [30] and [31] their compositions with the axisymmetric differential operator of potential theory were given. In particular, mapping properties of the generalized fractional calculus operators with the Gauss hypergeometric function as a kernel in $F_{p,\mu}$ and $F'_{p,\mu}$ were proved in [29]. We also mention the book [14], where special types of such H-transforms representable as compositions of the Erdélyi–Kober fractional integrals, were considered in $L_p(\mathbb{R}^+)(p \geq 1)$.

We also note that the relations similar to those in (1.11), (1.12), (1.13) and (1.14) for another Bessel-type integral transform were obtained in the subspace of locally integrable functions and in $F_{p,\mu}$ and $F'_{p,\mu}$ in [12], [13] and [6], respectively. The properties of such a transform, being introduced in [19], were also investigated in [26] and [5].

The paper is organized as follows. Section 2 deals with the properties of the function $\lambda_{\gamma,\sigma}^{(\beta)}(z)$ in (1.2) such as Mellin transform, asymptotic behavior near zero and infinity, fractional integration and differentiation of forms and . Section 3 contains preliminary results from the theory of spaces $F_{p,\mu}$ and $F'_{p,\mu}$ and mapping properties of operators (1.7), (1.8), (1.9) and (1.10) in these spaces. Section 4 is devoted to the modified Bessel-type integral transform (1.1) in the spaces $F_{p,\mu}$ and $F'_{p,\mu}$. 5 and 6 deal with compositions (1.11), (1.12) and (1.13), (1.14) in the space $F_{p,\mu}$. Such compositions in the space $F'_{p,\mu}$ are considered in 7 and 8.

The function $\lambda\gamma,\sigma(\beta)(z)$

First we note that the function $\lambda_\gamma^{(n)}(z)$ in (1.5) is expressed via $\lambda_{\gamma,\sigma}^{(\beta)}(z)$ in (1.2) when $\sigma=0$ and $\beta = n \in \mathbb{N}$:

$$\lambda_\gamma^{(n)}(z) = (2\pi)^{(n-1)/2} n^{-(n\gamma+1/2)} z^{\gamma n} \lambda_{\gamma,0}^{(n)}(z).$$

(2.1)

In particular, when n=1

$$\lambda_{\gamma,0}^{(1)}(z) = z^{-\gamma} \lambda_\gamma^{(1)}(z) = z^{-\gamma} e^{-z}.$$

(2.2)

Lemma 2.1:

Let $\beta \in \mathbb{R}^+, \sigma \in \mathbb{R}, \gamma \in \mathbb{C}$ with Re $(\gamma) > 1/\beta - 1$ and

$$\text{Re}(s) > \max[0, \beta \, \text{Re}(\gamma) + \sigma].$$

(2.3)

Then

$$(\mathfrak{M}\lambda_{\gamma,\sigma}^{(\beta)})(s) = \frac{\Gamma(s)\Gamma(-\gamma - \sigma/\beta + s/\beta)}{\Gamma(1 - (\sigma+1)/\beta + s/\beta)}.$$

(2.4)

Proof:

By (1.20) and (1.2) we have

$$
\begin{aligned}
(\mathfrak{M}\lambda_{\gamma,\sigma}^{(\beta)})(s) &= \frac{\beta}{\Gamma(\gamma+1-1/\beta)} \int_0^\infty y^{s-1}\,dy \int_1^\infty (t^\beta - 1)^{\gamma-1/\beta} t^\sigma e^{-yt}\,dt \\
&= \frac{\beta}{\Gamma(\gamma+1-1/\beta)} \int_1^\infty (t^\beta - 1)^{\gamma-1/\beta} t^\sigma \, dt \int_0^\infty y^{s-1} e^{-yt}\,dy \\
&= \frac{\beta\Gamma(s)}{\Gamma(\gamma+1-1/\beta)} \int_1^\infty (t^\beta - 1)^{\gamma-1/\beta} t^{\sigma-s}\,dt.
\end{aligned}
$$

Making change of the variable $t=u^{-1/\beta}$ and using the known relation between Beta and Gamma functions, we obtain

$$(\mathfrak{M}\lambda_{\gamma,\sigma}^{(\beta)})(s) = \frac{\Gamma(s)}{\Gamma(\gamma + 1 - 1/\beta)} \int_0^1 (1 - u)^{\gamma - 1/\beta} u^{-\gamma - (\sigma - s)/\beta - 1} \, du$$

$$= \frac{\Gamma(s)}{\Gamma(\gamma + 1 - 1/\beta)} B\left(\gamma + 1 - \frac{1}{\beta}, -\gamma - \frac{\sigma - s}{\beta}\right)$$

$$= \frac{\Gamma(s)\Gamma(-\gamma - \sigma/\beta + s/\beta)}{\Gamma(1 - (\sigma + 1)/\beta + s/\beta)}$$

and (2.4) is proved. The relation in (2.3) ensures the convergence of the integrals above.

Remark 2.2:

Using relation (2.4) and property (1.21) of the Mellin transform, we obtain the relation of form (1.22) for the modified Bessel-type integral transform (1.1)

$$(\mathfrak{M}L_{\gamma,\sigma}^{(\beta)}f)(s) = \frac{\Gamma(s)\Gamma(-\gamma - \sigma/\beta + s/\beta)}{\Gamma(1 - (\sigma + 1)/\beta + s/\beta)}(\mathfrak{M}f)(1 - s)$$

$$(2.5)$$

for "sufficiently good" function f(x). So, by (1.19) and (1.21), the kernel function $\lambda_{\gamma,\sigma}^{(\beta)}(z)$ can be expressed by the H-function

$$\lambda_{\gamma,\sigma}^{(\beta)}(z) = H_{1,2}^{2,0}\left[z \left| \begin{array}{l} (1 - (\sigma + 1)/\beta, 1/\beta) \\ (0, 1), (-\gamma - \sigma/\beta, 1/\beta) \end{array} \right.\right]$$

$$(2.6)$$

and the transform $L_{\gamma,\sigma}^{(\beta)}$ is the special case of the H-transform

$$(L_{\gamma,\sigma}^{(\beta)}f)(x) = \int_0^\infty H_{1,2}^{2,0}\left[xt \left| \begin{array}{l} (1 - (\sigma + 1)/\beta, 1/\beta) \\ (0, 1), (-\gamma - \sigma/\beta, 1/\beta) \end{array} \right.\right] f(t) \, dt.$$

$$(2.7)$$

The asymptotic behavior of $\lambda_{\gamma,\sigma}^{(\beta)}(z)$ near zero and infinity is given by

Lemma 2.3:

Let $\beta \in \mathbb{R}^+, \sigma \in \mathbb{R}$ and $\gamma \in \mathbb{C}$ with Re $(\gamma) > 1/\beta - 1$. Then

$$\lambda_{\gamma,\sigma}^{(\beta)}(z) \sim A \; (z \to 0) \quad with \; A = \frac{\Gamma(-\gamma - \sigma/\beta)}{\Gamma(1 - (\sigma + 1)/\beta)}, \tag{2.8}$$

provided that $Re(\gamma) < -\sigma/\beta$, and

$$\lambda_{\gamma,\sigma}^{(\beta)}(z) \sim Be^{-z}z^{-1-\gamma+1/\beta} \; (z \to \infty) \quad with \; B = \beta^{1+\gamma-1/\beta}. \tag{2.9}$$

Proof:

When z=0, we have

$$\lambda_{\gamma,\sigma}^{(\beta)}(0) = \frac{\beta}{\Gamma(\gamma + 1 - 1/\beta)} \int_1^\infty (t^\beta - 1)^{\gamma - 1/\beta} t^\sigma \, dt$$

which can be calculated as in the proof of Lemma 2.1, and

$$\lambda_{\gamma,\sigma}^{(\beta)}(0) = \frac{\Gamma(-\gamma - \sigma/\beta)}{\Gamma(1 - (\sigma + 1)/\beta)}.$$

Thus (2.8) is established. To prove (2.9), we rewrite $\lambda_{\gamma,\sigma}(\beta)(z)$ in the form

$$\lambda_{\gamma,\sigma}^{(\beta)}(z) = \frac{\beta e^{-z}}{\Gamma(\gamma + 1 - 1/\beta)} \int_0^\infty k(t)e^{-zt} \, dt \tag{2.10}$$

with

$$k(t) = [(1 + t)^\beta - 1]^{\gamma - 1/\beta}(1 + t)^\sigma.$$

Since

$$k(t) \sim \beta^{\gamma - 1/\beta} t^{\gamma - 1/\beta} \quad (t \to 0),$$

then by Watson's lemma (see, for example, [23])

$$\int_0^\infty k(t) e^{-zt} \, dt \sim \Gamma\left(\gamma + 1 - \frac{1}{\beta}\right) \beta^{\gamma - 1/\beta} z^{1/\beta - \gamma - 1} \quad (z \to \infty).$$

Hence

$$\lambda_{\gamma,\sigma}^{(\beta)}(z) \sim \beta^{\gamma - 1/\beta + 1} e^{-z} z^{1/\beta - \gamma - 1} \quad (z \to \infty)$$

and (2.9) holds. This completes the proof of Lemma 2.3.

Now we consider the right-sided fractional integration (1.9) and fractional differentiation (1.10) of the function $\lambda_{\gamma,\sigma}^{(\beta)}(x)$.

Lemma 2.4:

Let $\beta \in \mathbb{R}^+, \gamma \in \mathbb{C}, \sigma \in \mathbb{R}$ and Re $(\gamma) > 1/\beta - 1$. Then

$$(I_-^\alpha \lambda_{\gamma,\sigma}^{(\beta)})(x) = \lambda_{\gamma,\sigma-\alpha}^{(\beta)}(x), \tag{2.11}$$

$$(D_-^\alpha \lambda_{\gamma,\sigma}^{(\beta)})(x) = \lambda_{\gamma,\sigma+\alpha}^{(\beta)}(x). \tag{2.12}$$

Proof:

Using (1.9) and (1.2), changing the order of integration and applying [32, (5.20)], we have

$$(I_{-}^{\alpha} \lambda_{\gamma,\sigma}^{(\beta)})(x) = \frac{\beta}{\Gamma(\gamma + 1 - 1/\beta)} \int_{1}^{\infty} (t^{\beta} - 1)^{\gamma - 1/\beta} t^{\sigma} \, dt \frac{1}{\Gamma(\alpha)} \int_{x}^{\infty} (y - x)^{\alpha - 1} e^{-yt} \, dy$$

$$= \frac{\beta}{\Gamma(\gamma + 1 - 1/\beta)} \int_{1}^{\infty} (t^{\beta} - 1)^{\gamma - 1/\beta} t^{\sigma - \alpha} e^{-xt} \, dt = \lambda_{\gamma,\sigma-\alpha}^{(\beta)}(x),$$

which proves (2.11). Using (1.10) and (1.2), applying (2.11) with α being replaced by $1 - \{\alpha\}$ and taking the differentiation under the integral sign, we find

$$(D_{-}^{\alpha} \lambda_{\gamma,\sigma}^{(\beta)})(x) = \left(-\frac{d}{dx}\right)^{[\alpha]+1} (I_{-}^{1-\{\alpha\}} \lambda_{\gamma,\sigma}^{(\beta)})(x) = \left(-\frac{d}{dx}\right)^{[\alpha]+1} \lambda_{\gamma,\sigma-1+\{\alpha\}}^{\beta}(x)$$

$$= \frac{\beta}{\Gamma(\gamma + 1 - 1/\beta)} \int_{1}^{\infty} (t^{\beta} - 1)^{\gamma - 1/\beta} t^{\sigma - 1 + \{\alpha\}} \left(-\frac{d}{dx}\right)^{[\alpha]+1} e^{-xt} \, dt$$

$$= \frac{\beta}{\Gamma(\gamma + 1 - 1/\beta)} \int_{1}^{\infty} (t^{\beta} - 1)^{\gamma - 1/\beta} t^{\sigma + \alpha} e^{-xt} \, dt = \lambda_{\gamma,\sigma+\alpha}^{(\beta)}(x),$$

which completes the proof of Lemma 2.4.

Corollary 2.5:

Let $\beta \in \mathbb{R}^{+}, \gamma \in \mathbb{C}, \sigma \in \mathbb{R}$ and Re $(\gamma) > 1/\beta - 1$, then for $m \in \mathbb{N}$

$$\left(\frac{d}{dx}\right)^{m} \lambda_{\gamma,\sigma}^{(\beta)}(x) = (-1)^{m} \lambda_{\gamma,\sigma+m}^{(\beta)}(x).$$

$$(2.13)$$

By (2.1) and from Lemma 2.4 and Corollary 2.5, we obtain the corresponding results for the function $\lambda_{\gamma}^{(n)}(x)$ in (1.5).

Corollary 2.6:

Let $\alpha \in \mathbb{R}^+, \gamma \in \mathbb{C}, n \in \mathbb{N}$ and Re $(\gamma) > 1/n-1$. Then

$$(I_-^\alpha x^{-n\gamma} \lambda_\gamma^{(n)})(x) = (2\pi)^{(n-1)/2} n^{-(n\gamma+1/2)} \lambda_{\gamma,-\alpha}^{(n)}(x), \qquad (2.14)$$

$$(D_-^\alpha x^{-n\gamma} \lambda_\gamma^{(n)})(x) = (2\pi)^{(n-1)/2} n^{-(n\gamma+1/2)} \lambda_{\gamma,\alpha}^{(n)}(x). \qquad (2.15)$$

Corollary 2.7:

Let $\gamma \in \mathbb{C}, m, n \in \mathbb{N}$ and Re$(\gamma) > 1/n-1$, then

$$\left(\frac{d}{dx}\right)^m [x^{-n\gamma} \lambda_\gamma^{(n)}(x)] = (-1)^m (2\pi)^{(n-1)/2} n^{-(n\gamma+1/2)} \lambda_{\gamma,m}^{(n)}(x). \qquad (2.16)$$

Remark 2.8:

The relations similar to (2.11), (2.12) and (2.13) for another Bessel-type function were proved in [6].

SOME RESULTS FROM THE THEORY OF SPACES $F_{p,\mu}$ AND $F'_{p,\mu}$

We record here some results of McBride [21] and [22] and give mapping properties of the fractional calculus operators (1.7), (1.8), (1.9) and (1.10) in the spaces $F_{p,\mu}$ and $F'_{p,\mu}$.

Lemma 3.1:

McBride [21, Corollary 2.7].

The space $C^\infty_0(\mathbb{R}^+)$ of infinitely differentiable functions with compact supports in $\mathbb{R}^+$ is dense in $F_{p,\mu}$ for any $1 \le p \le \infty$ and $\mu \in \mathbb{C}$.

Lemma 3.2:

McBride [21, Theorems 2.11, 2.13].

Let $1 \leq p \leq \infty, \lambda, \mu \in \mathbb{C}$ and $m \in \mathbb{N}$.

i. The operator x^λ defined for $\varphi \in F_{p,\mu}$ by

$$(x^\lambda \varphi)(x) = x^\lambda \varphi(x) \tag{3.1}$$

is a homeomorphism of $F_{p,\mu}$ onto $F_{p,\mu,\mu+\lambda}$ with the inverse $x^{-\lambda}$.

ii. The operator D^m defined for $\varphi \in F_{p,\mu}$ by

$$(D^m \varphi)(x) = \frac{d^m \varphi(x)}{dx^m} \tag{3.2}$$

is a continuous linear mapping from $F_{p,\mu}$ into $F_{p,\mu-m}$. Further, D^m is a homeomorphism of $F_{p,\mu}$ onto $F_{p,\mu-m}$ if and only if

$$\text{Re}(\alpha) \neq \frac{1}{p} - k \quad (k = 0, 1, \ldots, m-1). \tag{3.3}$$

Lemma 3.3:

(McBride [21, Theorem 8.1, Corollory 8.2]). Let $1 \leq p \leq \infty, \lambda, \mu \in \mathbb{C}$ and let the function $k(x)$ be defined almost everywhere on $\mathbb{R}^+$ and satisfy

$$\int_0^\infty x^{\text{Re}(\mu)-1/p} |k(x)| \, dx < \infty. \tag{3.4}$$

If is an integral operator defined by

$$(\mathscr{K} \varphi)(x) = \int_0^\infty k(xt)\varphi(t) \, dt \quad (x > 0), \tag{3.5}$$

Then is a continuous linear mapping from $F_{p,\mu}$ into $F_{p,2/p-\mu-1}$.

Throughout the paper, for $p(1 \leq p \leq \infty)$ we denote $q(1 \leq q \leq \infty)$ by the relation

$$\frac{1}{p} + \frac{1}{q} = 1.$$

$$(3.6)$$

The mapping properties of the fractional integration operators in $F_{p,\mu}$ follow from McBride [21, Theorems 3.18, 3.23], namely.

Theorem 3.4:

Let $1 \leq p \leq \infty, \alpha \in \mathbb{R}^+, \mu \in \mathbb{C}$ and let I_{0+}^α and I_-^α be the operators (1.7) and (1.9), respectively.

i.   If Re $(\mu) > -1/q$, then I_{0+}^α is a continuous linear mapping from $F_{p,\mu}$ into $F_{p,\mu+\alpha}$.

ii. If Re $(\mu) > -1/p- \alpha$, then I_-^α is a continuous linear mapping from $F_{p,\mu}$ into $F_{p,\mu+\alpha}$.

According to (1.8) and (1.10) and from Lemma 3.2 and Theorem 3.4, we obtain mapping properties of fractional differentiation operators in $F_{p,\mu}$ of order $\alpha > (\alpha \notin \mathbb{N})$. When $\alpha = n \in \mathbb{N}$ we refer to Lemma 3.2(ii).

Theorem 3.5:

Let $1 \leq p \leq \infty, \alpha \in \mathbb{R}^+(\alpha \notin \mathbb{N}), \mu \in \mathbb{C}$ and letD0+$\alpha$and D$-\alpha$ be the operators (1.8) and (1.10), respectively. Then

i.   If Re $(\mu) > -1/q$, then D_{0+}^α is a continuous linear mapping from $F_{p,\mu}$ into$F_{p,\mu-\alpha}$.

ii. If Re $(\mu) < \{\alpha\text{-}1/q\}$, then D_-^{α} is a continuous linear mapping from $F_{p,\mu}$ into $F_{p,\mu-\alpha}$.

Let $1 \le p \le \infty, \alpha \in \mathbb{R}^+, \lambda, \mu \in \mathbb{C}$ and $m \in \mathbb{N}$. We denote by f an element of $F'_{p,\mu}$ and by $\langle f, \phi \rangle$ the value of f at a test function $\varphi \in F_{p,\mu}$. For any $f \in F'_{p,\mu}$ we denote by $x^{\lambda}, D^m, I_{0+}^{\alpha}, D_{0+}^{\alpha}, I_-^{\alpha}, D_-^{\alpha}$ the operators defined by (see [21] and [22]):

$$\langle x^{\lambda} f, \varphi \rangle = \langle f, x^{\lambda} \varphi \rangle \quad (\varphi \in F_{p,\mu-\lambda}), \tag{3.7}$$

$$\langle D^m f, \varphi \rangle = \langle f, (-1)^m D^m \varphi \rangle \quad (\varphi \in F_{p,\mu+m}), \tag{3.8}$$

$$\langle I_{0+}^{\alpha} f, \varphi \rangle = \langle f, I_-^{\alpha} \varphi \rangle \quad (\varphi \in F_{p,\mu-\alpha}), \tag{3.9}$$

$$\langle D_{0+}^{\alpha} f, \varphi \rangle = \langle f, D_-^{\alpha} \varphi \rangle \quad (\varphi \in F_{p,\mu+\alpha}), \tag{3.10}$$

$$\langle I_-^{\alpha} f, \varphi \rangle = \langle f, I_{0+}^{\alpha} \varphi \rangle \quad (\varphi \in F_{p,\mu-\alpha}), \tag{3.11}$$

$$\langle D_-^{\alpha} f, \varphi \rangle = \langle f, D_{0+}^{\alpha} \varphi \rangle \quad (\varphi \in F_{p,\mu+\alpha}). \tag{3.12}$$

Lemma 3.6: McBride [21, Theorem 2.22].

Let $1 \le p \le \infty, \lambda, \mu \in \mathbb{C}$ and $m \in \mathbb{N}$.

i. If operator x^{λ} then D_{0+}^{α} is a homeomorphism from $F'_{p,\mu}$ into.

ii. If operator D^m is a continuous linear mapping from $F'_{p,\mu}$ into $F'_{p,\mu+m}$. Further, D^m is a homeomorphism of $F'_{p,\mu}$ onto $F'_{p,\mu}$ if and only if.

$$\text{Re}(\mu) \ne \frac{1}{p} - k \quad (k = 1, \ldots, m). \tag{3.13}$$

From (3.9), (3.10), (3.11) and (3.12) and Theorem 3.4 and Theorem 3.5, we obtain the mapping properties of the fractional calculus operators in $F'_{p,\mu}$.

Theorem 3.7:

Let $1 \le p \le \infty, \alpha \in \mathbb{R}^+, \mu \in \mathbb{C}$ and let $I0+\alpha$ and $I-\alpha$ be operators defined in (3.9) and (3.11).

i.   If Re $(\mu) < 1/p$, then I^{α}_{0+} is a continuous linear mapping from $F'_{p,\mu}$ into $F'_{p,\mu-\alpha}$.

ii. If Re $(\mu) > \alpha - 1/q$, then I^{α}_- is a continuous linear mapping from $F'_{p,\mu}$ into $F'_{p,\mu-\alpha}$.

Theorem 3.8:

Let $1 \le p \le \infty, \alpha \in \mathbb{R}^+ (\alpha \notin \mathbb{C}), \mu \in \mathbb{C}$ and let D^{α}_{0+} and D^{α}_- be operators defined in (3.10) and (3.12)

i.   If Re $(\mu) < -[\alpha] - 1/p$, then D^{α}_{0+} is a continuous linear mapping from $F'_{p,\mu}$ into $F'_{p,\mu-\alpha}$.

ii. If Re $(\mu) > -\alpha - 1/q$, then D^{α}_- is a continuous linear mapping from $F'_{p,\mu}$ into $F'_{p,\mu-\alpha}$.

THE MODIFIED BESSEL-TYPE INTEGRAL TRANSFORM IN THE SPACES $F_{p,\mu}$ AND $F'_{p,\mu}$

Let us study the Bessel-type integral transform $L^{(\beta)}_{\gamma,\sigma}$ in the spaces $F_{p,\mu}$ and $F'_{p,\mu}$

$$\frac{1}{\beta} - 1 < \text{Re}(\gamma) < -\frac{\sigma}{\beta}, \quad \text{Re}(\mu) > -\frac{1}{q}. \tag{4.1}$$

Then the operator $L_{\gamma,\sigma}^{(\beta)}$ defined in (1.1) is a continuous linear mapping from $F_{p,\mu}$ and $F_{p,2/p-\mu-1}$

Proof:

We apply Lemma 3.3 to the operator $L_{\gamma,\sigma}^{(\beta)}$ in (1.1). The integral in (3.4) is equal to

$$\int_0^\infty x^{\mathrm{Re}(\mu)-1/p}\left|\lambda_{\gamma,\sigma}^{(\beta)}(x)\right|dx. \tag{4.2}$$

According to (2.8) and (2.9)

$$x^{\mathrm{Re}(\mu)-1/p}\left|\lambda_{\gamma,\sigma}^{(\beta)}(x)\right| \sim A x^{\mathrm{Re}(\mu)-1/p} \quad (x \to 0)$$

For $\mathrm{Re}(\gamma) < -\sigma/\beta$, and

$$x^{\mathrm{Re}(\mu)-1/p}\left|\lambda_{\gamma,\sigma}^{(\beta)}(x)\right| \sim B e^{-x} x^{\mathrm{Re}(\mu-\gamma)-1/p+1/\beta-1} \quad (x \to +\infty)$$

for $\mathrm{Re}(\gamma) > 1/\beta - 1$, where A and B are given (2.8) and (2.9). Hence the integral in (4.2) converges, and by Lemma 3.3 the operator $L_{\gamma,\sigma}^{(\beta)}$ is a continuous linear mapping from $F_{p,\mu}$ and $F_{p,2/p-\mu-1}$.

Theorem 4.2:

Let $1 \le p \le \infty, \beta \in \mathbb{R}^+, \gamma, \mu \in \mathbb{C}, \sigma \in \mathbb{R}(\sigma < \beta - 1)$ and let the conditions in (4:1) be satisfied. Then for $\varphi \in F_{p,\mu}$ and $\psi \in F_{q,2/q+\mu-1}$, there holds the formula of integration by parts

$$\int_0^\infty (L_{\gamma,\sigma}^{(\beta)}\psi)(x)\varphi(x)\,dx = \int_0^\infty \psi(x)(L_{\gamma,\sigma}^{(\beta)}\varphi)(x)\,dx. \tag{4.3}$$

Proof:

Relation (4.3) for "sufficiently good" function φ and ψ is verified by interchanging the order of integration. To show that (4.3) holds for $\varphi \in F_{p,\mu}$ and $\psi \in F_{q,2/q+\mu-1}$, it is sufficient to prove that both sides of (4.3) represent bounded linear functional on $L_\mu^p \times L_{2/q+\mu-1}^q$, where

$$L_\mu^p = \{\varphi: x^{-\mu}\varphi(x) \in L^p(\mathbb{R}^+)\}$$

with the norm

$$\|\varphi\|_{p,\mu} = \left(\int_0^\infty |x^{-\mu}\varphi(x)|^p \, dx\right)^{1/p}$$

(see [21,22]). In fact, from Hölder's inequality and Theorem 4.1 we have

$$\left(\int_0^\infty (L_{\gamma,\sigma}^{(\beta)}\psi)(x)\varphi(x)\,dx\right) \leqslant \int_0^\infty |x^{-\mu}\varphi(x)||x^\mu(L_{\gamma,\sigma}^{(\beta)}\psi)(x)|\,dx$$

$$\leqslant \left(\int_0^\infty |x^{-\mu}\varphi(x)|^p\,dx\right)^{1/p}\left(\int_0^\infty |x^\mu L_{\gamma,\sigma}^{(\beta)}\psi(x)|^q\,dx\right)^{1/q}$$

$$= \|\varphi\|_{p,\mu}\|L_{\gamma,\sigma}^{(\beta)}\psi\|_{q,-\mu} \leqslant k\|\varphi\|_{p,\mu}\|\psi\|_{q,2/q+\mu-1},$$

where k is a positive constant. Therefore the left-hand side of (4.3) represents a bounded linear functional on $L_\mu^p \times L_{2/q+\mu-1}^q$ as, similarly, does the right-hand side of (4.3). Thus the theorem is proved.

Due to Theorem 4.2 we define the operator $L_{\gamma,\sigma}^{(\beta)}$ f for f $\in F'_{p,\mu}(1 \leq p \leq \infty; \mu \in \mathbb{C})$ by

$$\langle L_{\gamma,\sigma}^{(\beta)}f, \varphi\rangle = \langle f, L_{\gamma,\sigma}^{(\beta)}\varphi\rangle \quad (\varphi \in F_{p,2/p-\mu-1}). \tag{4.4}$$

Then from Theorem 4.1 we obtain the mapping property of the operator $L_{\gamma,\sigma}^{(\beta)}$ in the space $F'_{p,\mu}$.

Theorem 4.3:

Let $1 \le p \le \infty, \beta \in \mathbb{R}^+, \gamma, \mu \in \mathbb{C}, \sigma \in \mathbb{R}(\sigma < \beta - 1)$ and

$$\frac{1}{\beta} - 1 < \mathrm{Re}(\gamma) < -\frac{\sigma}{\beta}, \quad \mathrm{Re}(\mu) < \frac{1}{p}. \tag{4.5}$$

Then the operator $L_{\gamma,\sigma}^{(\beta)}$ defined by (4.4) is a continuous linear mapping from $F'_{p,\mu}$ into $F_{p,2/p-\mu-1}$

COMPOSITIONS OF $L_{\gamma,\sigma}^{(\beta)}$ AND I0+A, D0+A IN $F_{p,\mu}$

We prove relations and for $\varphi \in F_{p,\mu}$. We begin with the first one.

Theorem 5.1:

Let $1 \le p \le \infty, \alpha, \beta \in \mathbb{R}^+, \gamma, \mu \in \mathbb{C}, \sigma \in \mathbb{R}(\sigma < \beta - 1)$ and let $L_{\gamma,\sigma}^{(\beta)}$ and I_{0+}^{α} operators (1.1) and (1.7). If the conditions in (4.1) are satisfied, then for $\varphi \in F_{p,\mu}$ relation (1.11) holds, i.e.,

$$L_{\gamma,\sigma}^{(\beta)} I_{0+}^{\alpha} \varphi = x^{-\alpha} L_{\gamma,\sigma-\alpha}^{(\beta)} \varphi. \tag{5.1}$$

Proof:

Let $\varphi \in C_0^{\infty}(\mathbb{R}^+)$. In view of (1.1), (1.2) and (1.7), the interchange of the order of integration yields

$$(L_{\gamma,\sigma}^{(\beta)} I_{0+}^{\alpha} \varphi)(x) = \frac{1}{\Gamma(\alpha)} \int_0^{\infty} \varphi(t)\,\mathrm{d}t \int_t^{\infty} (y-t)^{\alpha-1} \lambda_{\gamma,\sigma}^{(\beta)}(xy)\,\mathrm{d}y$$

$$= \frac{\beta}{\Gamma(\alpha)\Gamma(\gamma+1-1/\beta)} \int_0^{\infty} \varphi(t)\,\mathrm{d}t \int_1^{\infty} (\tau^{\beta}-1)^{\gamma-1/\beta} \tau^{\sigma}\,\mathrm{d}\tau$$

$$\times \int_t^{\infty} (y-t)^{\alpha-1} e^{-xy\tau}\,\mathrm{d}y.$$

Applying [32, (5.20)], we obtain

$$(L_{\gamma,\sigma}^{(\beta)} I_{0+}^{\alpha} \varphi)(x) = \frac{\beta}{\Gamma(\gamma + 1 - 1/\beta)} \int_0^\infty \varphi(t)\, dt \int_1^\infty (\tau^\beta - 1)^{\gamma - 1/\beta} \tau^\sigma (x\tau)^{-\alpha} e^{-xt\tau}\, d\tau$$

$$= x^{-\alpha} (L_{\gamma,\sigma-\alpha}^{(\beta)} \varphi)(x),$$

and (5.1) is proved for $\varphi \in C_0^\infty(\mathbb{R}^+)$. According to (4.1), Theorem 4.1 and Theorem 3.4) and Lemma 3.2(i), the operators in both sides of (5.1) are continuous linear mapping from $F_{p,\mu}$ into $F_{p,2/p-\mu-\alpha-1}$. Therefore, by Lemma 3.1 and the Banach theorem, (5.1) holds for $\varphi \in F_{p,\mu}$.

Theorem 5.2:

Let $1 \le p \le \infty, \alpha, \beta \in \mathbb{R}^+, \gamma, \mu \in \mathbb{C}, \sigma \in \mathbb{R}(\sigma < \beta - \alpha - 1)$ and let $L_{\gamma,\sigma}^{(\beta)}$ and D_{0+}^α be operators (1.1) and (1.8). If

$$\mathrm{Re}(\mu) > \alpha - \frac{1}{q}, \quad \frac{1}{\beta} - 1 < \mathrm{Re}(\gamma) < -\frac{\sigma + \alpha}{\beta}, \tag{5.2}$$

then for $\varphi \in F_{p,\mu}$ relation (1.12) holds, i.e.,

$$L_{\gamma,\sigma}^{(\beta)} D_{0+}^\alpha \varphi = x^\alpha L_{\gamma,\sigma+\alpha}^{(\beta)} \varphi. \tag{5.3}$$

Proof:

Let $\alpha > 0 (\alpha \notin \mathbb{N}), \varphi \in C_0^\infty(\mathbb{R}^+), n = [\alpha] + 1$ and let $\phi_{n-a} \equiv I_{0+}^{n-a} \phi$. According to (1.1), (1.2) and (1.3), integrating by part n times and using relation (2.13), we have

$$(L_{\gamma,\sigma}^{(\beta)} D_{0+}^{\alpha}\varphi)(x) = \int_0^{\infty} \lambda_{\gamma,\sigma}^{(\beta)}(xt)\varphi_{n-\alpha}^{(n)}(t)\,dt$$

$$= \lambda_{\gamma,\sigma}^{(\beta)}(xt)\varphi_{n-\alpha}^{(n-1)}(t)\Big|_{t=0}^{\infty} + x\int_0^{\infty} \lambda_{\gamma,\sigma+1}^{(\beta)}(xt)\varphi_{n-\alpha}^{(n-1)}(t)\,dt$$

$$= \ldots$$

$$= \sum_{k=0}^{n-1} x^k \lambda_{\gamma,\sigma+k}^{(\beta)}(xt)(I_{0+}^{n-\alpha}\varphi)^{(n-1-k)}(t)\Big|_{t=0}^{\infty} + x^n(L_{\gamma,\sigma+n}^{(\beta)}I_{0+}^{n-\alpha}\varphi)(x).$$

$$(5.4)$$

When $\alpha = m \in \mathbb{N}$, this formula takes the form

$$(L_{\gamma,\sigma}^{(\beta)} D_{0+}^{\alpha}\varphi)(x) = \sum_{k=0}^{m-1} x^k \lambda_{\gamma,\sigma+k}^{(\beta)}(xt)\varphi^{(m-1-k)}(t)\Big|_{t=0}^{\infty} + x^m(L_{\gamma,\sigma+m}^{(\beta)}\varphi)(x). \qquad (5.5)$$

It is directly verified that the functions $(I_{0+}^{n-\alpha}\phi)^{(n-1-k)}$ and $\phi^{(m-1-k)}$ belong to $C_0^{\infty}(\mathbb{R}^+)$ and therefore

$$(I_{0+}^{n-\alpha}\varphi)^{(n-1-k)}(0) = (I_{0+}^{n-\alpha}\varphi)^{(n-1-k)}(\infty) = 0 \quad (k = 0, 1, \ldots, n-1)$$

and

$$\varphi^{(m-1-k)}(0) = \varphi^{(m-1-k)}(\infty) = 0 \quad (k = 0, 1, \ldots, m-1).$$

Then from (5.4) we obtain

$$(L_{\gamma,\sigma}^{(\beta)} D_{0+}^{\alpha}\varphi)(x) = x^n(L_{\gamma,\sigma+n}^{(\beta)}I_{0+}^{n-\alpha}\varphi)(x).$$

Applying (5.1) with σ replaced by $\sigma+n$ and α by $n-\alpha$, we have for $\alpha \notin \mathbb{N}$,

$$(L_{\gamma,\sigma}^{(\beta)} D_{0+}^{\alpha}\varphi)(x) = x^{\alpha}(L_{\gamma,\sigma+\alpha}^{(\beta)}\varphi)(x)$$

which is just (5.3) for $\varphi \in C_0^\infty(\mathbb{R}^+)$. For the case $\alpha = m \in \mathbb{N}$, (5.3) is also implied by (5.5). According to (5.2), Theorem 4.1 and Theorem 3.5) and Lemma 3.2(i), the operators on both the sides of (5.3) are continuous linear mappings from $F_{p,\mu}$ into $F_{p,2/p-\mu-\alpha-1}$ and hence (5.3) holds for $\varphi \in F_{p,\mu}$ in accordance with Lemma 3.1 and the Banach theorem.

Corollary 5.3: *Let* $1 \le p \le \infty, \beta \in \mathbb{R}^+, \gamma, \mu \in \mathbb{C}, m \in \mathbb{N}$ *and* $\sigma \in \mathbb{R}(\sigma < \beta - m - 1)$ *and let*

$$\mathrm{Re}(\mu) > m - \frac{1}{q}, \quad \frac{1}{\beta} - 1 < \mathrm{Re}(\gamma) < -\frac{\sigma + m}{\beta}, \tag{5.6}$$

then for $\varphi \in F_{p,\mu}$

$$L_{\gamma,\sigma}^{(\beta)} D^m \varphi = x^m L_{\gamma,\sigma+m}^{(\beta)} \varphi. \tag{5.7}$$

Corollary 5.4: *Let* $1 \le p \le \infty, \beta \in \mathbb{R}^+, \gamma, \mu \in \mathbb{C}, \sigma \in \mathbb{R}(\sigma < \beta - 2)$, *and let*

$$\mathrm{Re}(\mu) > \frac{1}{p}, \quad \frac{1}{\beta} - 1 < \mathrm{Re}(\gamma) < -\frac{\sigma + 1}{\beta}, \tag{5.8}$$

then for $\varphi \in F_{p,\mu}$

$$L_{\gamma,\sigma}^{(\beta)} D\varphi = x L_{\gamma,\sigma+1}^{(\beta)} \varphi. \tag{5.9}$$

Corollary 5.5: *Let* $1 \le p \le \infty, \beta \in \mathbb{R}^+, \gamma, \mu \in \mathbb{C}, \sigma \in \mathbb{R}(\sigma < \beta - 3)$ *and let*

$$\mathrm{Re}(\mu) > 1 + \frac{1}{p}, \quad \frac{1}{\beta} - 1 < \mathrm{Re}(\gamma) < -\frac{\sigma + 2}{\beta}, \tag{5.10}$$

then for $\varphi \in F_{p,\mu}$

$$L_{\gamma,\sigma}^{(\beta)} D^2 \varphi = x^2 L_{\gamma,\sigma+2}^{(\beta)} \varphi. \tag{5.11}$$

Remark 5.6:

The relation similar to (5.1), (5.3), (5.7), (5.9) and (5.11) for another Bessel-type integral transform was proved in [12] and [13] on the subspace of the space of locally integrable functions and in [6] on $F_{p,\mu}$.

COMPOSITION OF $L_{\gamma,\sigma}^{(\beta)}$ AND I_-^α, D_-^α IN $F_{p,\mu}$

Relations (1.13) and (1.14) for $\varphi \in F_{p,\mu}$ can be obtained analogously as stated in the following without proof:

Theorem 6.1:

Let $1 \le p \le \infty, \alpha, \beta \in \mathbb{R}^+, \gamma, \mu \in \mathbb{C}, \sigma \in \mathbb{R}(\sigma < \beta - 1)$ and let $L_{\gamma,\sigma}^{(\beta)}$ and I_-^α operators (1.1) and (1.9). If

$$\mathrm{Re}(\mu) > \alpha - \frac{1}{q}, \quad \frac{1}{\beta} - 1 < \mathrm{Re}(\gamma) < -\frac{\sigma}{\beta}, \tag{6.1}$$

then for $\varphi \in F_{p,\mu}$ relation (1.13) holds, i.e.,

$$I_-^\alpha L_{\gamma,\sigma}^{(\beta)} \varphi = L_{\gamma,\sigma-\alpha}^{(\beta)} x^{-\alpha} \varphi. \tag{6.2}$$

Theorem 6.2:

Let $1 \le p \le \infty, \alpha, \beta \in \mathbb{R}^+, \gamma, \mu \in \mathbb{C}, \sigma \in \mathbb{R}(\sigma < \beta - \alpha - 1)$ and let $L_{\gamma,\sigma}^{(\beta)}$ and D_-^α be operators (1.1) and (1.10). If

$$\text{Re}(\mu) > \begin{cases} -\{\alpha\} + \dfrac{1}{p} & (\alpha \notin \mathbb{N}), \\ -\dfrac{1}{q} & (\alpha \in \mathbb{N}), \end{cases} \quad \dfrac{1}{\beta} - 1 < \text{Re}(\gamma) < -\dfrac{\sigma + \alpha}{\beta}, \tag{6.3}$$

then for $\varphi \in F_{p,\mu}$ relation (1.14) holds, i.e.,

$$D_-^{\alpha} L_{\gamma,\sigma}^{(\beta)} \varphi = L_{\gamma,\sigma+\alpha}^{(\beta)} x^{\alpha} \varphi. \tag{6.4}$$

Corollary 6.3:

Let $1 \le p \le \infty, \beta \in \mathbb{R}^+, \gamma, \mu \in \mathbb{C}, m \in \mathbb{N}, \sigma \in \mathbb{R}(\sigma < \beta - m - 1)$ and

$$\text{Re}(\mu) > -\dfrac{1}{q}, \quad \dfrac{1}{\beta} - 1 < \text{Re}(\gamma) < -\dfrac{\sigma + m}{\beta}, \tag{6.5}$$

then for $\varphi \in F_{p,\mu}$

$$D^m L_{\gamma,\sigma}^{(\beta)} \varphi = (-1)^m L_{\gamma,\sigma+m}^{(\beta)} x^m \varphi. \tag{6.6}$$

Corollary 6.4:

Let $1 \le p \le \infty, \beta \in \mathbb{R}^+, \gamma, \mu \in \mathbb{C}, \sigma \in \mathbb{R}(\sigma < \beta - 2)$ and

$$\text{Re}(\mu) > -\dfrac{1}{q}, \quad \dfrac{1}{\beta} - 1 < \text{Re}(\gamma) < -\dfrac{\sigma + 1}{\beta}, \tag{6.7}$$

then for $\varphi \in F_{p,\mu}$

$$DL_{\gamma,\sigma}^{(\beta)} \varphi = -L_{\gamma,\sigma+1}^{(\beta)} x \varphi. \tag{6.8}$$

Corollary 6.5:

Let $1 \leq p \leq \infty, \beta \in \mathbb{R}^+, \gamma, \mu \in \mathbb{C}, \sigma \in \mathbb{R}(\sigma < \beta - 3)$ and

$$Re(\mu) > -\frac{1}{q}, \quad \frac{1}{\beta} - 1 < Re(\gamma) < -\frac{\sigma + 2}{\beta}, \tag{6.9}$$

then for $\varphi \in F_{p,\mu}$

$$D^2 L_{\gamma,\sigma}^{(\beta)} \varphi = L_{\gamma,\sigma+2}^{(\beta)} x^2 \varphi. \tag{6.10}$$

Remark 6.6:

The relations similar to (6.2), (6.4), (6.6), (6.8) and (6.10) for another Bessel-type integral transform were obtained in [12] and [13] and [7] in the space of locally integrable functions and in $F_{p,\mu}$, respectively.

COMPOSITION OF $L_{\gamma,\sigma}^{(\beta)}$ AND $I_{0+}^\alpha, D_{0+}^\alpha$ IN $F'_{p,\mu}$

We prove formulas (1.11) and (1.12) for $f \in F'_{p,\mu}$. We begin from the first one.

Theorem 7.1:

Let $1 \leq p \leq \infty, \alpha, \beta \in \mathbb{R}^+, \gamma, \mu \in \mathbb{C}, \sigma \in \mathbb{R}(\sigma < \beta - 1)$ and let $L_{\gamma,\sigma}^{(\beta)}$ and I_{0+}^α the operators defined in (4.3) and (3.9). If the conditions in (4.5) are satisfied, then for $f \in F'_{p,\mu}$ relation (1.11) holds, i.e.,

$$L_{\gamma,\sigma}^{(\beta)} I_{0+}^\alpha f = x^{-\alpha} L_{\gamma,\sigma-\alpha}^{(\beta)} f. \tag{7.1}$$

Proof:

According to Theorems 4.3, 3.7(i) and Lemma 3.6(i) the operators on both sides of (7.1) are continuous linear mappings from $F'_{p,\mu}$ into $F'_{p,2/p-\mu+\alpha-1}$. Therefore to establish (7.1), it is sufficient to prove the relation

$$\langle L_{\gamma,\sigma}^{(\beta)} I_{0+}^{\alpha} f, \varphi \rangle = \langle x^{-\alpha} L_{\gamma,\sigma-\alpha}^{(\beta)} f, \varphi \rangle$$

(7.2)

for $\varphi \in F_{p,2/p-\mu+\alpha-1}$. Applying (4.4), (3.9), (6.2), (4.4) and (3.7), we have

$$\langle L_{\gamma,\sigma}^{(\beta)} I_{0+}^{\alpha} f, \varphi \rangle = \langle I_{0+}^{\alpha} f, L_{\gamma,\sigma}^{(\beta)} \varphi \rangle = \langle f, I_{-}^{\alpha} L_{\gamma,\sigma}^{(\beta)} \varphi \rangle$$
$$= \langle f, L_{\gamma,\sigma-\alpha}^{(\beta)} x^{-\alpha} \varphi \rangle = \langle L_{\gamma,\sigma-\alpha}^{(\beta)} f, x^{-\alpha} \varphi \rangle = \langle x^{-\alpha} L_{\gamma,\sigma-\alpha}^{(\beta)} f, \varphi \rangle$$

(7.3)

and (7.2) is proved. We note that in the third equality in (7.3) we have used formula (6.2), where condition (6.1) (with μ replaced by $2/p-\mu+\alpha-1$) is equivalent to condition (4.5).

Theorem 7.2:

Let $1 \le p \le \infty, \beta \in \mathbb{R}^+, \gamma, \mu \in \mathbb{C}, \sigma \in \mathbb{R}(\sigma < \beta - \alpha - 1)$ and let $L_{\gamma,\sigma}^{(\beta)}$ and D_{0+}^{α} be the operators defined in (4.4) and (3.10).
(i) If $\alpha \notin \mathbb{N}$ and

$$\text{Re}(\mu) < -[\alpha] - \frac{1}{q}, \quad \frac{1}{\beta} - 1 < \text{Re}(\gamma) < -\frac{\sigma + \alpha}{\beta},$$

(7.4)

then for $f \in F'_{p,\mu}$ relation (1.12) holds, i.e.,

$$L^{(\beta)}_{\gamma,\sigma} D^{\alpha}_{0+} f = x^{\alpha} L^{(\beta)}_{\gamma,\sigma+\alpha} f. \tag{7.5}$$

(ii) If $\alpha = m \in \mathbb{N}$ and

$$\mathrm{Re}(\mu) < -m + \frac{1}{p}, \quad \frac{1}{\beta} - 1 < \mathrm{Re}(\gamma) < -\frac{\sigma + m}{\beta}, \tag{7.6}$$

then for $f \in F'_{p,\mu}$ relation (1.12) holds, i.e.,

$$L^{(\beta)}_{\gamma,\sigma} D^m f = x^m L^{(\beta)}_{\gamma,\sigma+m} f. \tag{7.7}$$

Proof:

The operations on both sides of (7.5) are continuous linear mappings from $F'_{p,\mu}$ into $F_{p,2/p-\mu-\alpha-1}$ by Theorems 3.8 (i), 4.3 and Lemma 3.6, provided that the conditions in (7.4) are satisfied. Therefore, as in the proof of Theorem 7.1, it is sufficient to show that

$$\langle L^{(\beta)}_{\gamma,\sigma} D^{\alpha}_{0+} f, \varphi \rangle = \langle x^{\alpha} L^{(\beta)}_{\gamma,\sigma+\alpha} f, \varphi \rangle \tag{7.8}$$

for $\varphi \in F_{p,2/p-\mu-\alpha-1}$. If $\alpha \notin \mathbb{N}$, then applying (4.4), (3.10), (6.4), (4.4) and (3.7), we have

$$\langle L^{(\beta)}_{\gamma,\sigma} D^{\alpha}_{0+} f, \varphi \rangle = \langle D^{\alpha}_{0+} f, L^{(\beta)}_{\gamma,\sigma} \varphi \rangle = \langle f, D^{\alpha}_{-} L^{(\beta)}_{\gamma,\sigma} \varphi \rangle$$

$$= \langle f, L^{(\beta)}_{\gamma,\sigma+\alpha} x^{\alpha} \varphi \rangle = \langle L^{(\beta)}_{\gamma,\sigma+\alpha} f, x^{\alpha} \varphi \rangle = \langle x^{\alpha} L^{(\beta)}_{\gamma,\sigma+\alpha} f, \varphi \rangle. \tag{7.9}$$

If $\alpha = m \in \mathbb{N}$ then the application of (4.4), (3.8), (6.6), (4.4) and (3.7) deduce for

$$\langle L_{\gamma,\sigma}^{(\beta)} D^m f, \varphi \rangle = \langle D^m f, L_{\gamma,\sigma}^{(\beta)} \varphi \rangle = \langle f, (-1)^m D^m L_{\gamma,\sigma}^{(\beta)} \varphi \rangle$$

$$= \langle f, L_{\gamma,\sigma+m}^{(\beta)} x^m \varphi \rangle = \langle L_{\gamma,\sigma+m}^{(\beta)} f, x^m \varphi \rangle = \langle x^m L_{\gamma,\sigma+m}^{(\beta)} f, \varphi \rangle. \quad (7.10)$$

From (7.9) and (7.10) we arrive at (7.8) and theorem is proved. We only note that in the third equalities in (7.9) and (7.10) we have used formulas (6.4) and (6.6), where condition (6.3) (with μ replaced by 2/p$-\mu-\alpha-1$) is equivalent to (7.4).

Corollary 7.3:

If $1 \le p \le \infty, \beta \in \mathbb{R}^+, \gamma, \mu \in \mathbb{C}, \gamma, \sigma \in \mathbb{R}(\sigma < \beta - 2)$ and

$$\text{Re}(\mu) < -\frac{1}{q}, \quad \frac{1}{\beta} - 1 < \text{Re}(\gamma) < -\frac{\sigma + 1}{\beta},$$

$$(7.11)$$

then for $f \in F'_{p,\mu}$

$$L_{\gamma,\sigma}^{(\beta)} Df = x L_{\gamma,\sigma+1}^{(\beta)} f.$$

$$(7.12)$$

Corollary 7.4: If $1 \le p \le \infty, \beta \in \mathbb{R}^+, \gamma, \mu \in \mathbb{C}, \sigma \in \mathbb{R}(\sigma < \beta - 3)$ *and*

$$\text{Re}(\mu) < -2 + \frac{1}{p}, \quad \frac{1}{\beta} - 1 < \text{Re}(\gamma) < -\frac{\sigma + 2}{\beta},$$

$$(7.13)$$

then for $f \in F'_{p,\mu}$

$$L_{\gamma,\sigma}^{(\beta)} D^2 f = x^2 L_{\gamma,\sigma+2}^{(\beta)} f.$$

$$(7.14)$$

Remark 7.5:

The relations similar to (7.1), (7.5), (7.7), (7.12) and (7.14) for another Bessel-type integral transform were established in [12] and [13] and [6] in the subspace of locally integrable functions and in $F'_{p,\mu}$, respectively.

COMPOSITION OF $L^{(\beta)}_{\gamma,\sigma}$ AND $I^{\alpha}_{-}, D^{\alpha}_{-}$ IN $F'_{p,\mu}$

Finally, we give the formulas relating to (1.13) and (1.14) for $f \in F'_{p,\mu}$. The proofs are omitted.

Theorem 8.1:

Let $1 \le p \le \infty, \alpha, \beta \in \mathbb{R}^{+}, \gamma, \mu \in \mathbb{C}, \sigma \in \mathbb{R}(\sigma < \beta - 1)$ and let $L^{(\beta)}_{\gamma,\sigma}$ and I^{α}_{-} be the operators defined in (4.4) and (3.11). If

$$Re(\mu) < \frac{1}{p} - \alpha, \qquad \frac{1}{\beta} - 1 < Re(\gamma) < -\frac{\sigma}{\beta}, \tag{8.1}$$

then for $f \in F'_{p,\mu}$ relation (1.13) holds, i.e.,

$$I^{\alpha}_{-} L^{(\beta)}_{\gamma,\sigma} f = L^{(\beta)}_{\gamma,\sigma-\alpha} x^{-\alpha} f. \tag{8.2}$$

Theorem 8.2:

Let $1 \le p \le \infty, \alpha, \beta \in \mathbb{R}^{+}, \gamma, \mu \in \mathbb{C}, \sigma \in \mathbb{R}(\sigma < \beta - \alpha - 1)$ and let $L^{(\beta)}_{\gamma,\sigma}$ and D^{α}_{-} the operators defined in (4.4) and (3.12). If

$$Re(\mu) < \frac{1}{p}, \qquad \frac{1}{\beta} - 1 < Re(\gamma) < -\frac{\sigma + \alpha}{\beta}, \tag{8.3}$$

then for $f \in F'_{p,\mu}$ the type of relation (1.14) holds, i.e.,

$$D^{\alpha}_{-} L^{(\beta)}_{\gamma,\sigma} f = L^{(\beta)}_{\gamma,\sigma+\alpha} x^{\alpha} f.$$

$$(8.4)$$

Corollary 8.3:

Let $1 \le p \le \infty, \beta \in \mathbb{R}^{+}, \gamma, \mu \in \mathbb{C}, \sigma \in \mathbb{R}(\sigma < \beta - 2)$ and

$$\mathrm{Re}(\mu) < \frac{1}{p}, \quad \frac{1}{\beta} - 1 < \mathrm{Re}(\gamma) < -\frac{\sigma+1}{\beta},$$

$$(8.5)$$

then for $f \in F'_{p,\mu}$

$$DL^{(\beta)}_{\gamma,\sigma} f = L^{(\beta)}_{\gamma,\sigma+1} x f.$$

$$(8.6)$$

Corollary 8.4:

Let $1 \le p \le \infty, \beta \in \mathbb{R}^{+}, \gamma, \mu \in \mathbb{C}, \sigma \in \mathbb{R}(\sigma < \beta - 3)$ and

$$\mathrm{Re}(\mu) < \frac{1}{p}, \quad \frac{1}{\beta} - 1 < \mathrm{Re}(\gamma) < -\frac{\sigma+2}{\beta},$$

$$(8.7)$$

then for $f \in F'_{p,\mu}$

$$D^{2} L^{(\beta)}_{\gamma,\sigma} f = L^{(\beta)}_{\gamma,\sigma+2} x^{2} f.$$

$$(8.8)$$

Remark 8.5:

The relations similar to (8.2), (8.5) and (8.8) and (8.10) for another Bessel-type integral transform were obtained in [12] and [13] and [6] in the subspace of locally integrable functions and in $F'_{p,\mu}$, respectively.

ACKNOWLEDGMENTS

The present investigation was supported, in part, by the Belarusian Fundamental Research Fund, and by the Science Research Promotion Fund from the Japan Private School Promotion Foundation.

REFERENCES

1. J.J. Betancor, C.J. Diaz, Boundedness and range of H-transformation on certain weighted L_p spaces, Serdica 20 (1994) 269–297.

2. Erd e' lyi, A.C. McBride, Fractional integrals of distributions, SIAM J. Math. Anal. 1 (1970) 547–557.

3. Erd e' lyi, W. Magnus, F. Oberhettinger, F.G. Tricomi, Higher Transcendental Functions, Vol. II, McGraw-Hill, New York, 1953.

4. H.-J. Glaeske, A.A. Kilbas, M. Saigo, S.A. Shlapakov, $L_{v,r}$; theory of integral transformations with the H-function in the kernel, Dokl. Akad. Nauk Belarusi 41=2 (1997) 10 –15 (in Russian).

5. H.-J. Glaeske, A.A. Kilbas, Bessel-type integral transforms on $L_{v,r}$-spaces, Results Math. 34 (1998) 320–329.

6. A.A. Kilbas, B. Bonilla, M. Rivero, J. Rodrigues, J. Trujillo, Compositions of Bessel type integral transform with fractional operators on spaces $F_{p,\mu}$ and $F'_{p,\mu}$, Frac. Calcul. Appl. Anal. 1 (1998) 135–150.

7. A.A. Kilbas, B. Bonilla, J. Rodrigues, J. Trujillo, M. Rivero, Compositions of fractional integrals and derivatives with Bessel type function and solution of differential and integral equations, Dokl. Nats. Akad. Nauk Belarusi 42=2 (1998), 25 –29 (in Russian).

8. A.A. Kilbas, M. Saigo, On generalized fractional integration operators with Fox's H-function on spaces $F_{p,\mu}$ and $F'_{p,\mu}$, Integral Transforms Spec. Funct. 4 (1996) 103–114.

9. A.A. Kilbas, M. Saigo, S.A. Shlapakov, Integral transforms with Fox's H-function in spaces of summable functions, Integral Transforms Spec. Funct. 1 (1993) 87–103.

10. A.A. Kilbas, M. Saigo, S.A. Shlapakov, Integral transforms with Fox's H-function in $L_{v,r}$-spaces, Fukuoka Univ. Sci. Rep. 23 (1993) 9–31.

11. A.A. Kilbas, M. Saigo, S.A. Shlapakov, Integral transforms with Fox's H-function in $L_{v,r}$-spaces, II, Fukuoka Univ. Sci. Rep. 24 (1994) 13–38.

12. A.A. Kilbas, S.A. Shlapakov, On a Bessel-type integral transformation and its compositions with integral and differential operators, Dokl. Akad. Nauk Belarus 37=4 (1993) 10 –14 (in Russian).

13. A.A. Kilbas, S.A. Shlapakov, On the composition of a Bessel-type integral operator with operators of fractional integro-differentiation and the solution of differential equations, Differential Equations 30 (1994) 235–246.

14. V.S. Kiryakova, Generalized Fractional Calculus and Applications, Res. Notes Math., Vol. 301, Pitman, San Francisco, 1994.

15. E. Krätzel, Eine Verallgemeinerung der Laplace- und Meijer-Transformation, Wiss. Z. Friedrich-Schiller-Univ. Jena=Thuringen 14 (1965) 369–381.

16. E. Krätzel, Die Faltung der L-Transformation, Wiss. Z. Friedrich-Schiller-Univ. Jena=Thuringen 14 (1965) 383–390.

17. E. Krätzel, Bemerkungen zur Meijer-Transformation und Anwendungen, Math. Nachr. 30 (1965) 327–334.

18. E. Krätzel, Differentiationssätze der L-Transformation und Differentialgleichungen nach dem Operator $(d / dt)t^{1/n-v}(t^{1-1/n}d / dt)^{n-1}t^{v+1-2n}$ Math. Nachr. 35 (1967) 105–114.

19. E. Krätzel, Integral transformations of Bessel-type, Generalized Functions and Operational Calculus (Proc. Conf., Varna, 1975), Bulgar. Acad. Sci., Soa, 1979, pp. 148–155.

20. A.C. McBride, A theory of fractional integrals of generalized functions, SIAM J. Math. Anal. 6 (1975) 583–599.

21. A.C. McBride, Fractional Calculus and Integral Transforms of Generalized Functions, Res. Notes Math., Vol. 31, Pitman, San Francisco, 1979.

22. A.C. McBride, Fractional powers of a class of ordinary differential operators, Proc. London Math. Soc., Ser. 3 45 (1982) 519–546.

23. F.W.J. Olver, Asymptotics and Special Functions, Academic Press, New York, 1974.

24. A.P. Prudnikov, Yu.A. Brychkov, O.I. Marichev, Integrals and Series, Vol. 3, More Special Functions, Gordon & Breach, New York, 1990.

25. R.K. Raina, M. Saigo, Fractional calculus operators involving Fox's H-function in spaces $F_{p,\mu}$ and $F'_{p,\mu}$, Recent Advances in Fractional Calculus, Global Publ., Souk Rapids (Minnesota), 1993, pp. 219 –229.

26. J. Rodriguez, J.J. Trujillo, M. Rivero, Operational fractional calculus of Kratzel integral transformation, Differential Equations (Xanthi, 1987), Lecture Notes in Pure Applied Mathematics, Vol. 118, Dekker, New York, 1989, pp. 613– 620.

27. P.G. Rooney, A technique for studying the boundedness and extendability of certain types of operators, Cand. J. Math. 25 (1973) 1090–1102.

28. P.G. Rooney, On integral transformations with G-function kernels, Proc. Royal Soc. Edinburgh, Sec. A 93 (1982=83) 265–297.

29. M. Saigo, H.-J. Glaeske, Fractional calculus operators involving the Gauss function in spaces $F_{p,\mu}$ and $F'_{p,\mu}$, Math. Nachr. 147 (1990) 285–306.

30. M. Saigo, A.A. Kilbas, Compositions of generalized fractional calculus operators with axisymmetric differential operator of potential theory on spaces $F_{p,\mu}$ and $F'_{p,\mu}$, Boundary Value Problems, Special Functions and Fractional Calculus, Beloruss Gos. Univ, Minsk, 1996, pp. 335 –350.

31. M. Saigo, R.K. Raina, A.A. Kilbas, On generalized fractional calculus operators and their compositions with axisymmetric differential operator of the potential theory on spaces $F_{p,\mu}$ and $F'_{p,\mu}$, Fukuoka Univ. Sci. Rep. 23 (1993) 133–154.

32. S.G. Samko, A.A. Kilbas, O.I. Marichev, Fractional Integrals and Derivatives, Theory and Applications, Gordon & Breach, Yverdon, 1993.

33. S.A. Shlapakov, M. Saigo, A.A. Kilbas, On inversion of H-transform in $L_{v,r}$-spaces, Internat. J. Math. Math. Sci. 21 (1998) 713–722.

34. H.M. Srivastava, K.C. Gupta, S.P. Goyal, The H-functions of One and Two Variables with Applications, South Asia Publ., New Delhi, 1982.

CITATION

H.-J. Glaeske, Anatoly A. Kilbas, Megumi Saigo, A modified Bessel-type integral transform and its compositions with fractional calculus operators on spaces $F_{p,\mu}$ and $F'_{p,\mu}$, Journal of Computational and Applied Mathematics, Volume 118, Issues 1–2, 1 June 2000, Pages 151-168, ISSN 0377-0427, http://dx.doi.org/10.1016/S0377-0427(00)00286-7.

Existence Theorem for a Nonlinear Functional Integral Equation and an Initial Value Problem of Fractional Order in $L_1(R_+)$

Ibrahim Abouelfarag Ibrahim[1, 2], Tarek S. Amer[1, 3], and Yasser M. Aboessa[1, 4]

[1]Mathematics Department, Faculty of Science and Education, Taif University, Al-Khurmah Branch, Taif, KSA
[2]Mathematics Department, Faculty of Science, Suez Canal University, Ismailia, Egypt
[3]Mathematics Department, Faculty of Science, Tanta University, Tanta, Egypt
[4]Mathematics Department, Faculty of Science, Al-Azhar University, Cairo, Egypt

ABSTRACT

The aim of this paper is to study the existence of integrable solutions of a nonlinear functional integral equation in the space of Lebesgue integrable functions on unbounded interval, $L_1(R_+)$. As an application we deduce the existence of solution of an initial value problem of fractional order that be studied only on a bounded interval. The main tools used are Schauder fixed point theorem, measure of weak noncompactness, superposition operator and fractional calculus.

INTRODUCTION

The class of functional integral equations of various types plays very important role in numerous mathematical research areas. An interesting feature of functional integral equations is its role in the study of many problems of functional differential Equations [1-4].

In this work we study the solvability of the following initial value problem

$$\frac{dy(t)}{dt} = \int_0^t k_1(t,s) f\left(s, D^\beta y(s)\right) ds,$$

$$t \in R_+, \beta \in (0,1], y(0) = y_0.$$

$$(1)$$

where $D^\beta y$ denotes the fractional derivative of order β of y with $\beta \in (0,1]$. Such initial value problem of arbitrary order (1) was investigated in [5-7]. To achieve this goal, let us consider the integral equation

$$x(t) = \int_0^t k_1(t,s) f\left(s, \int_0^s k_2(s,\theta) x(\phi(\theta)) d\theta\right) ds,$$

$$t \in R_+$$

$$(2)$$

which is different from that studied in [2].

Section 2 contains some basic results. Our main result will be given in Section 3. Solvability of the considered initial value problem will be discussed in Section 4.

BASIC CONCEPTS

This section is devoted to recall some notations and known results that will be needed in the sequel.

If A is a Lebesgue measurable subset of the set of real numbers R then we use the symbol meas.(A) to denote the Lebesgue measure of A. Let $L_1(A)$ be the space of all real functions defined and Lebesgue measurable on the set A. If $x \in L_1(A)$ then the norm of x is defined as:

$$\|x\| = \|x\|_{L_1(A)} = \int_A |x(t)| \, dt.$$

when $A = R_+[0,\infty)$, we will write L_1 instead of $L_1(R_+)$.

The Superposition Operator

An important operator called the superposition operator can be investigated in the theories of differential integral and functional equations [4,8-10]. It can be defined as follows:

Definition 1

Assume that $f : I \times R \rightarrow R$ satisfies Carathéodory conditions, that is it is measurable in t for any x and continuous in x for almost all t where $t \in I, x \in R$. Then for every measurable function x on the interval I we assign the function:

$$(Fx)(t) = f(t, x(t)), t \in I.$$

The operator F defined in this way is called the superposition operator generated by the function f.

Carathéodory [11] gave the first contribution to the theory of the superposition operator and proved its measurability according to the measurability of f.

We state the following result giving the necessary and sufficient condition so that the superposition operator F generated by f will map continuously L_1 into itself [12].

Theorem 2

Let f satisfy the conditions in Definition 1. The superposition operator F generated by the function f maps continuously the space L_1 into itself if and only if:

$$|f(t,x)| \le a(t) + b|x|,$$

for all $t \in I$ and $x \in R$, where α is a function that belongs to L_1 and b is a nonnegative constant.

It is known that a real valued continuous function is measurable and that the converse is not necessarily true. However, for the converse we have the following results due to Dragoni [13].

Theorem 3

Let I be a bounded interval and $f : I \times R \to R$ be a function satisfying Caratheodory conditions. Then for each $\varepsilon > 0$ there exists a closed subset D_ε of the interval I such that $\text{meas.}(I / D_\varepsilon) < \varepsilon$ and $f|_{D_\varepsilon \times R}$ is continuous.

Volterra Integral Operator

We proceed by recalling some basic facts concerning the linear Volterra integral operator in the Lebesgue space L_1. Suppose $K : \Delta \to R$ is a given function which is measurable with respect to both variables where

$$\Delta = \{(t,s) : 0 \le s \le t < \infty\}.$$

For an arbitrary function $x \in L_1$ define Volterra integral operator as follows:

$$(Kx)(t) = \int_0^t k(t,s)x(s)ds, t \in R_+.$$

It is well known that if $K : L_1 \to L_1$ then it is continuous [4, 9].

In general, it is rather difficult to find necessary and sufficient conditions for the function $k(t,s)$ guaranteeing that the integral operator K transforms the space L_1 into itself. Some special cases of this problem were discussed in [4, 14]. In this direction we state the next result [15]:

Theorem 4

Let k be measurable on Δ and such that

$$\operatorname{ess\,sup}_{s \geq 0} \int_s^\infty \left| k(t,s) \right| ds < \infty.$$

Then the Volterra integral operator K generated by k maps (continuously) the space $L_1 = L_1(R_+)$ into itself and the norm $\|K\|$ of this operator is majorized by the number

$$\operatorname{ess\,sup}_{s \geq 0} \int_s^\infty | k(t,s) | \, ds.$$

Observe that if D is a nonempty and measurable subset of R_+ then we can also consider the linear Volterra integral operator associated with the Lebesgue space $L_1(D)$

Namely If $x \in L_1(D)$ where D is a nonempty and measurable subset of R_+ then we extend x to the whole half axis R_+ by putting $x(t) = 0$ for $t \in R_+ / D$. Then we can treat K in the usual way. When the operator K transforms $L_1(D)$ into itself its norm will be denoted by $\|K\|_D$.

Measures of Weak Noncompactness

Let us assume that E is an infinite dimensional Banach space with the norm $\|.\|$ and the zero elements θ. Denote by m_E the family of all non-empty and bounded subsets of E and by n_E^W its subfamily consisting of all relatively weakly compact sets. The symbol $\bar{X}^W$ stands for the weak closure of a set X and the symbol ConvX will denote the convex closed hull (with respect to the norm topology) of a set X. We denote by B(x,r) the ball centered at x and of radius r. We write B_r instead of $B(\theta,r)$. In what follows we accept the following definition [16]

Definition 5

A function $\mu : m_E \rightarrow R_+$ is said to be a measure of weak noncompactness if it satisfies the following conditions: The Family

1) The family $\mathrm{Ker}\,\mu = \{X \in m_E : \mu(X) = 0\}$ is nonempty and is nonempty and ker $\mu \subset \bar{n}_E^W$

2) $X \subset Y \Rightarrow \mu(X) \leq \mu(Y)$.

3) $\mu(ConvX) = \mu(X)$.

4) $\mu(\lambda X + (1-\lambda)Y) \leq \lambda\mu(X) + (1-\lambda)\mu(Y)$, for $\lambda \in [0,1]$.

5) $X_n \in m_E, X_n = \bar{X}_n^W$ and $X_{n+1} \subset X_n$ for $n = 1,2,\cdots$

And if $\lim_{n\to\infty}\mu(X_n) = 0$ then the intersection is nonempty $X_\infty = \bigcap_\infty^\infty X_n$

The family Ker μ is said to be the kernel of the measure of weak non-compactness μ. Let us observe that the intersection set X_∞ from 5) belongs to ker μ. Indeed, since $\mu(X_\infty) \le \mu(X_n)$ for every n then we have that $\mu(X_\infty) = 0$.

We can construct a useful measure of weak noncompactness in the space L_1 that based on the following criterion for weak noncompact-ness due to Dieudonné [17, 18].

Theorem 6

A bounded set X is relatively weakly compact in L_1 if and only if the following two conditions are satisfied:

a) for any $\varepsilon > 0$ there exists $\delta > 0$ such that if meas. $(D) \le \delta$ Then
$$\int_D | x(t) | \, dt \le \varepsilon \quad \text{for all, } x \in X.$$

b) for any $\varepsilon > 0$ there is T>0 such that $\int_T^\infty | x(t) | \, dt \le \varepsilon$ for any , $x \in X$.

Now, for a nonempty and bounded subset X of the space L_1 let us define:

$$\mu(X) = c(X) + d(X),$$

(3)

where

$$c(X) = \lim_{\varepsilon \to 0} \left\{ \sup_{x \in X} \left\{ \sup \left\{ \int_D |x(t)| \, dt : D \subset R_+, \text{meas } D \le \varepsilon \right\} \right\} \right\},$$

and

$$d(X) = \lim_{T \to \infty} \left\{ \sup \left\{ \int_{T}^{\infty} |x(t)| dt : x \in X \right\} \right\}.$$

It can be shown [17] that the function μ is a measure of weak noncompactness in the space L_1 such that $\beta(X) \le \mu(X) \le 2\mu(X)$, for any $X \in m_{L_1}$, where β denotes the De Blasi measure of weak noncompactness in L_1. Moreover, $\mu(\beta_r) = 2r$.

In our approach we will need the following fixed point theorem due to Schauder.

Theorem 7

Let C be a nonempty, convex, closed, and bounded subset of a Banach space E. Let $H : C \to C$ be a completely continuous mapping. Then H has at least one fixed point in C.

Fractional Calculus

The definitions of both differential operator and the integral operator of fractional order are stated as follows [19, 20].

Definition 8

Let $f \in L_1, \alpha \in R_+$. The RiemmanLiouville (R-L) fractional integral of the function f(t) of order α is defined as

$$I_a^\alpha f(t) = \int_a^t \frac{(t-s)^{\alpha-1}}{\Gamma(\alpha)} f(s)\,ds,$$

$$\alpha > 0, a \le t \le b.$$

Definition 9

Let g(t) be an absolutely continuous function on [a,b]. Then the fractional derivative of order $\alpha \in (0,1]$ of g(t) is defined as

$$D_a^\alpha g(t) = I_a^{1-\alpha} Dg(t),$$
$$D = (d/dt).$$

We state here some results concerning the above mentioned operators:

1) Let $f, Df \in L$ and $\alpha, \beta \in (0,1]$, then

    i. then $\,_a^\alpha I_a^\beta f(t) = I_a^{\alpha+\beta} f(t).$

    ii. $DI_a^\alpha f(t) = I_a^\alpha Df(t),$ when $f(a) = 0.$

2) The operator I_a^α maps L_1 into itself continuously.

EXISTENCE THEOREM

Consider the integral Equation (2) and let H denotes the operator determined by the right hand side of this equation, i.e.,

$$(Hx)(t) = \int_0^t k_1(t,s) f\left(s, \int_0^s k_2(s,\theta) x(\varphi(\theta)) d\theta \right) ds, \tag{4}$$

where $t \in R_+$ In fact the operator H can be written as the product $H = K_1 F_1 K_2 F_2$ of the linear Volterra operator

$$(K_i x)(t) = \int_0^t k_i(t,s) x(s) ds, i = 1, 2.$$

and the superposition operator

$$(Fx)(t) = f(t, x(t)), t \in R_+.$$

Therefore Equation (4) can be written as:

$$x = Hx = K_1 F_1 K_2 F_2 x(\varphi).$$

(5)

To establish our main result concerning existence of an integrable solution of Equation (2) we impose suitable conditions on the functions involved in that equation. Namely we assume

1) The functions $f : R_+ \times R \to R$ satisfy the Caratheodory conditions and there exist functions $a \in L_1$ and constants $b > 0$ such that
$$|f(t,x)| \leq a(t) + b|x|$$

   holds for all $(t,x) \in R^+ \times R$.

2) The functions $k_i : R_+ \times R_+ \to R$ satisfy the Caratheodory conditions and the linear Volterra operators K_1, K_2 associated with k_1, k_2 map L_1 into itself.

3) $\varphi : R_+ \to R_+$ is increasing, absolutely continuous and there exists a constant M>0 such that $\varphi'(t) \geq M$ a.e. on R_+.

4) $b \|K_1\| \|K_2\| M^{-1} < 1$.

Now we can state our main result in the next theorem.

Theorem 10

Under the above assumptions the Equation (2) has at least one solution $x \in L_1$.

Proof. Since H is a nonlinear operator defined by Equation (5), then based on assumptions i) and ii) if $x \in L_1.$, then $Hx \in L_1$. Moreover, from Equation (5), and noting that K_1, K_2 according to our assumptions are indeed bounded, we have

$$\|Hx\| = \|K_1 F_1 K_2 F_2 x(\varphi)\| \le \|K_1\| \|F_1 K_2 F_2 x(\varphi)\|$$

$$\le \|K_1\| \int_0^\infty \left| f\left(s, \int_0^s k_2(s,\theta) x(\varphi(\theta)) d\theta \right) \right| ds$$

$$\le \|K_1\| \int_0^\infty \left\{ a(s) + b \left| \int_0^s k_2(s,\theta) x(\varphi(\theta)) d\theta \right| \right\} ds$$

$$\le \|K_1\| \|a\| + b \|K_1\| \|K_2 x(\varphi)\|$$

$$\le \|K_1\| \|a\| + b \|K_1\| \|K_2\| \int_0^\infty |x(\varphi(\theta))| d\theta$$

$$\le \|K_1\| \|a\| + b \|K_1\| \|K_2\| \int_0^\infty |x(\varphi(\theta))| \frac{\varphi'(\theta)}{M} d\theta$$

$$\le \|K_1\| \|a\| + b \|K_1\| \|K_2\| M^{-1} \int_0^\infty |x(u)| du$$

$$\le \|K_1\| \|a\| + b \|K_1\| \|K_2\| M^{-1} \|x\|.$$

The above estimate shows that the operator H maps B_r into itself, where

$$r = \left(\|K_1\| \|a\| (1-b) \|K_1\| \|K_2\| M^{-1} \right)^{-1}.$$

Moreover, according to Theorem 2, we deduce that the operator H is continuous on the space L_1.

Next, to prove that H is a contraction, let X be a nonempty subset of B_r. Fix $\varepsilon > 0$ and take a measurable subset $D \subset R_+$ such that meas $D \leq \varepsilon$. Then for any $x \in X$, we get

$$\int_D \left| (Hx)(t) \right| dt$$

$$= \int_D \left| \left(K_1 F K_2 x(\varphi) \right)(t) \right| dt \leq \|K_1\|_D \left\| F_1 K_2 x(\varphi) \right\|_{L_1(D)}$$

$$\leq \|K_1\|_D \int_D \left| f\left(s, \int_0^s k_2(s,\theta) x(\varphi(\theta)) d\theta \right) \right| ds,$$

$$\leq \|K_1\|_D \int_D \left| a(s) + b \int_0^s k_2(s,\theta) x(\varphi(\theta)) d\theta \right| ds$$

$$\leq \|K_1\|_D \int_D \left| a(t) \right| dt + b \|K_1\|_D \int_D \left| (K_2 x(\varphi))(t) \right| dt$$

$$\leq \|K_1\|_D \int_D \left| a(t) \right| dt + b \|K_1\|_D \left\| K_2 x(\varphi) \right\|_{L_1(D)}$$

$$\leq \|K_1\|_D \int_D \left| a(t) \right| dt + b \|K_1\|_D \|K_2\|_D \left\| x(\varphi) \right\|_{L_1(D)}$$

$$\leq \|K_1\|_D \int_D \left| a(t) \right| dt + \frac{b \|K_1\|_D \|K_2\|_D}{M} \int_{\varphi(D)} \left| x(u) \right| du$$

where the symbol $\|.\|_D$ denotes the operator norm acting from the space $L_1(D)$ into itself. Also in the above calculation we used the fact that

$a(t) \geq 0$ for $t \in R_+$. From the absolute continuity of the function φ and the obvious equality

$$\lim_{\varepsilon \to 0} \left\{ \sup \left\{ \int_D a(t)\,dt : D \subset R_+, \text{meas.}D \leq \varepsilon \right\} \right\} = 0.$$

and using Theorem 6 we obtain

$$c(HX) \leq \left(\frac{b\|K_1\|_D \|K_2\|_D}{M} \right) c(X).$$

(6)

Furthermore, fixing T>0 we can deduce that

$$\int_T^\infty |(Hx)(t)|\,dt$$

$$= \|K_1 FK_2 x(\varphi)\|_{L_1[T,\infty)} = \int_T^\infty |(K_1 FK_2 x(\varphi))(t)|\,dt$$

$$\leq K_{1T} \int_T^\infty |a(t)|\,d + \frac{bK_{1T}K_{2T}}{M} \int_{\varphi(T)}^\infty |x(u)|\,du,$$

where the symbol $\|\ \|_T$ denotes the operator norm acting from the space $L_1[T,\infty)$ into itself. Now according to the fact that the set consisting of one element is weakly compact, by using Theorem 6 and the formula

$$d(X) = \lim_{T \to \infty} \left\{ \sup \left\{ \int_T^\infty |x(t)|\,dt : x \in X \right\} \right\}.$$

and since $\lim_{T\to\infty}\varphi(T) = \infty$, we get

$$d\left(HX\right) \le \left(\frac{b\|K_1\|\|K_2\|}{M}\right)d\left(X\right).$$

(7)

According to Equation (3), combining (6) and (7), we get

$$\mu\left(HX\right) \le \left(\frac{b\|K_1\|\|K_2\|}{M}\right)\mu\left(X\right)$$

(8)

Put $q = (b\|K_1\|\|K_2\|)/M$. Clearly, according to assumption iv) q<1. Consider the sequence of sets $\{B_r^n\}$, where $B_r^1 = \text{Conv}(GB_r), B_r^2 = \text{Conv}(GB_r^1)$ and so on. Obviously this sequence is decreasing i.e. $B_r^{n+1} \subset B_r^n$ for $n = 1, 2, \cdots$. Moreover, $B_r^1 \subset B_r$. Apart from this, all sets belonging to this sequence are closed and convex, so weakly closed. On the other hand in view of inequality (8) we have

$$\mu\left(B_r^n\right) \le q^n \mu\left(B_r\right),$$

which yields that $\lim_{n\to\infty}\mu(B_r^n) = 0$.

Consequently, by axiom 5) of Definition 5 we infer that the set $Y = \bigcap_{n=1}^{\infty} B_r^n$ is nonempty, closed, convex and weakly compact (in view of $\mu(Y) \overset{=}{} 0$). Moreover, $GY \subset Y$.

In the sequel we show that the set GY is relatively compact in the set L_1.

To do this let us take an arbitrary sequence $\{y_n\} \subset Y$ and fix arbitrarily a number $\varepsilon > 0$. Since Y is weakly compact, in view of Theorem 6 we deduce that there exists $T > 0$ such that for any natural number n the following inequality is satisfied

$$\int_T^\infty |y_n(t)| dt \leq \frac{\varepsilon}{4}.$$

(9)

To apply the classical Schauder fixed point theorem, we need to prove that the set HY is relatively compact in L_1. For this aim let us consider the functions f(t,x) on the set $[0, T]$ and the functions $k_i(t,s)$ on the set

$$[0,T] \times [0,T] (i = 1,2).$$

In view of Theorem 3 we can find a closed subset D_ε of the interval $[0, T]$ such that meas.$(D_\varepsilon^c) \leq \varepsilon$ (where $D_\varepsilon^c = [0, T] \setminus D_\varepsilon$) and such that the functions $f|_{D_\varepsilon \times R}$ and $K_i|_{D_\varepsilon \times [0,T]}$ $(i = 1,2)$ are continuous. Hence we infer that $K_i|_{D_\varepsilon \times [0,T]}$ $(i = 1,2)$ are uniformly continuous.

In what follows we show that $\{y_n\}$ is an equicontinuous on D_ε, for that let us take arbitrarily $t_1, t_2 \in D_\varepsilon$. Without loss of generality we can assume that $t_1 < t_2$. Then, keeping in mind our assumptions, for an arbitrary fixed $n \in N$ we obtain:

$$\left| Hy_n(t_2) - Hy_n(t_1) \right|$$

$$\leq \left| \int_0^{t_2} k_1(t_2,s) f\left(s, \int_0^s k_2(s,\theta) y_n(\varphi(\theta)) d\theta \right) ds \right.$$

$$\left. - \int_0^{t_1} k_1(t_1,s) f\left(s, \int_0^s k_2(s,\theta) y_n(\varphi(\theta)) d\theta \right) ds \right|$$

$$\leq \omega^T\left(k_1, |t_2 - t_1| \right) \int_0^T \left\{ a(s) + b\bar{k}_2 \int_0^s \left| y_n(\varphi(\theta)) \right| d\theta \right\} ds$$

$$+ \bar{k}_1 \int_{t_1}^{t_1} \left\{ a(s) + b\bar{k}_2 \int_0^s \left| y_n(\varphi(\theta)) \right| d\theta \right\} ds$$

where $\omega^T(k_1,.)$ denotes the modulus of continuity of the function k_1 on the set $D_\varepsilon \times [0,T]$ and

$$\bar{k}_i = \max\left\{ \left| k_i(t,s) \right| : (t,s) \in D_\varepsilon \times [0,T], i = 1,2 \right\}.$$

$$\left| Hy_n(t_2) - Hy_n(t_1) \right|$$

$$\leq \omega^T\left(k_1, |t_2 - t_1| \right) \left\{ \int_0^T a(s) ds + b\bar{k}_2 \int_0^T \int_0^s \left| y_n(\varphi(\theta)) \right| d\theta ds \right\}$$

$$+ \bar{k}_1 \int_{t_1}^{t_2} a(s) ds + b\bar{k}_1 \bar{k}_2 \int_{t_1}^{t_2} \int_0^s \left| y_n(\varphi(\theta)) \right| d\theta ds$$

By rearranging the order of double integrations, we get

$$\left| Hy_n\left(t_2\right) - Hy_n\left(t_1\right)\right| \le \omega^I\left(k_1, t_2 - t_1\right)$$

$$\cdot \left\{ \int_0^T a\left(s\right)ds + b\overline{k}_2 \int_0^T \int_\theta^T \left| y_n\left(\varphi(\theta)\right)\right| ds d\theta \right\}$$

$$+ \overline{k}_1 \int_{t_1}^{t_2} a\left(s\right)ds + b\overline{k}_1\overline{k}_2 \int_0^{t_1} \int_{t_1}^{t_2} \left| y_n\left(\varphi(\theta)\right)\right| ds d\theta$$

$$+ b_1\overline{k}_1\overline{k}_2 \int_{t_1}^{t_2} \int_\theta^{t_2} \left| y_n\left(\varphi(\theta)\right)\right| ds d\theta$$

$$\le \omega^I\left(k^1, t_2 - t^1\right)\left\{ \int_0^T a\left(s\right)ds + b\overline{k}_2 T \int_0^T \left| y_n\left(\varphi(\theta)\right)\right| d\theta \right\}$$

$$+ \overline{k}_1 \int_{t_1}^{t_2} a\left(s\right)ds + b\overline{k}_1\overline{k}_2\left(t_2 - t_1\right) \int_0^{t_1} \left| y_n\left(\varphi(\theta)\right)\right| d\theta$$

$$+ b\overline{k}_1\overline{k}_2\left(t_2 - t_1\right) \int_{t_1}^{t_2} \left| y_n\left(\varphi(\theta)\right)\right| d\theta$$

From the above estimate and the consideration of the fact that $Y \subset B_r$ we obtain

$$\left| H y_n'(t_2) - H y_n'(t_1) \right| \le \omega^T (k_1, t_2 - t_1)$$

$$\cdot \left\{ \int_0^\infty a(s)\, ds + b \overline{k}_2 T \int_0^\infty \left| v_n(\varphi(\theta)) \right| d\theta \right\}$$

$$+ \overline{k}_1 \int_{t_1}^{t_2} a(s)\, ds + b \overline{k}_1 \overline{k}_2 (t_2 - t_1) \int_0^{t_1} \left| v_n(\varphi(\theta)) \right| d\theta$$

$$+ b \overline{k}_1 \overline{k}_2 (t_2 - t_1) \int_{t_1}^{t_2} \left| v_n(\varphi(\theta)) \right| d\theta$$

$$\le \omega^T (k_1, t_2 - t_1) \left\{ a + \frac{b T \overline{k}_2}{M} r \right\} + \overline{k}_1 \int_{t_1}^{t_2} a(s)\, ds$$

$$+ \frac{b \overline{k}_1 \overline{k}_2 (t_2 - t_1)}{M} r.$$

Now, utilizing the fact that the sequence $\{y_n\}$ is weakly compact and

taking into account Theorem 6 we can show that the number $\int^{t_2} a(s)ds$ is arbitrarily small provided the number $(t_2 - t_1)$ is taken to be sufficiently small (it is a consequence of the fact that a one element set is weakly compact in L_1).

Furthermore,

$$\left\| \left(Hy'_n \right)(t) \right\|$$

$$= \left| \int_0^t k_1(t,s) f\left(s, \int_0^s k_2(s,\theta) y_n(\varphi(\theta)) d\theta \right) ds \right|$$

$$\leq \int_0^t |k_1(t,s)| \left\{ a(s) + b \int_0^s |k_2(s,\theta)| |v_n(\varphi(\theta))| d\theta \right\} ds$$

$$\leq \overline{k}_1 \int_0^t a(s) ds + b\overline{k}_1 \overline{k}_2 \int_0^t \int_\theta^t |v_n(\varphi(\theta))| ds d\theta$$

$$\leq \overline{k}_1 \int_0^t a(s) ds + b\overline{k}_1 \overline{k}_2 \int_0^t |v_n(\varphi(\theta))| \int_\theta^t ds d\theta$$

$$\leq \overline{k}_1 \int_0^t a(s) ds + b\overline{k}_1 \overline{k}_2 \int_0^t |v_n(\varphi(\theta))| (t-\theta) d\theta$$

$$\leq \overline{k}_1 \int_0^t a(s) ds + b\overline{k}_1 \overline{k}_2 T \int_0^t |v_n(\varphi(\theta))| d\theta$$

$$\leq \overline{k}_1 \int_0^t a(s) ds + b\overline{k}_1 \overline{k}_2 T \int_0^t |v_n(\varphi(\theta))| \frac{\varphi'(\theta)}{M} d\theta$$

$$\leq \overline{k}_1 \int_0^t a(s) ds + \frac{b\overline{k}_1 \overline{k}_2 T}{M} \int_0^{\varphi(t)} |v_n(u)| du$$

Hence

$$\left\| \left(Hy_n \right)(t) \right\| \leq \overline{k}_1 \|a\| + b\overline{k}_1 \overline{k}_2 T M^{-1} r.$$

Hence consequently the sequence {Hy$_n$} is a sequence of uniformly bounded and equicontinuous functions on D$_\varepsilon$. Hence, in view of Ascoli-Arzela theorem we deduce that the sequence $\{Hy_n\}$ is relatively compact subset in the space $C(D_\varepsilon)$.

Further observe that the above reasoning does not depend on the choice of ε. Thus we can construct a sequence $\{D_p\}$ of closed subsets of the interval [0, T] such that meas.$(D_p^c) \to 0$ as $P \to \infty$ and such that the sequence $\{Hy_n\}$ is relatively compact in every space

$C(D_p)$. Passing to subsequences if necessary we can assume that $\{Hy_n\}$ is a Cauchy sequence in each space $C(D_p)$, for $p = 1, 2, \cdots$.

In what follows, utilizing the fact that the set HY is weakly compact, let us choose a number $\delta > 0$ such that for each closed subset D_δ of the interval [0, T] such that meas.$(D_\delta^c) \leq \delta$ we have

$$\int_{D_\delta^c} |(Hy)(t)| \, dt \leq \frac{\varepsilon}{4} \tag{10}$$

for any $y \in Y$.

Keeping in mind the fact that the sequence $\{Hy_n\}$ is a Cauchy sequence in each space $C(D_p)$ we can choose a natural number P_0 such that meas.$(D_{P_0}^c) \leq \delta$ and for arbitrary natural numbers $n, m \geq p_0$ the following inequality holds

$$\left| (Hy_n)(t) - (Hy_m)(t) \right| \leq \frac{\varepsilon}{4 \text{meas.}\left(D_{P_0} \right)}$$

for any $t \in D_{P0}$. Obviously without loss of generality we can assume that meas.$(D_{P0}) > 0$.

Now, using the above facts and (10) we obtain

$$\int_0^T \left|\left(Hy_n\right)(t)-\left(Hy_m\right)(t)\right| dt$$

$$= \int_{D_{p0}} \left|\left(Hy_m\right)(t)-\left(Hy_m\right)(t)\right| dt$$

$$+ \int_{D_{p0}^c} \left|\left(Hy_m\right)(t)-\left(Hy_m\right)(t)\right| dt$$

$$\leq \frac{\varepsilon \mathrm{meas.}\left(D_{p^0}\right)}{4\mathrm{meas.}\left(D_{p^0}\right)}$$

$$+ \int_{D_{p0}^c} \left\{\left|\left(Hy_m\right)(t)\right|+\left|\left(Hy_m\right)(t)\right|\right\} dt = \frac{3\varepsilon}{4}.$$

$$(11)$$

Finally, from (10) and (11) we get

$$\left\|Hy_m - Hy_n\right\| = \int_0^\infty \left|\left(Hy_n\right)(t)-\left(Hy_m\right)(t)\right| dt \leq \varepsilon,$$

which means that $\{Hy_n\}$ is a Cauchy sequence in the space $L_1 = L_1(R_+)$. Hence we conclude that the set HY is relativelycompact in this space.

In the last step of the proof let us consider the set $Y_0 = \mathrm{Conv}(HY)$. In view of the Mazur theorem we infer that the set Y_0 is compact in the space L_1. Moreover, we have that the operator H transforms continuously the set Y_0 into itself. Thus the classical Schauder fixed point principle gives that H has at least one fixed point. This proves that there exists at least one $x \in L_1$ that solves Equation (4).

NONLINEAR EQUATION OF CONVOLUTION TYPE

Assume that $k : R_+ \to R$ is an integrable function. For an arbitrary function $x \in L_1$ set

$$(Kx)(t) = \int_0^t k(t-s)x(s)\,ds, t \in R_+ .$$

This operator K is a linear integral operator of convolution type and maps L_1 into itself continuously.

Now, consider the following condition

$$(v)\, k : R_+ \to R \text{ and } k \in L_1.$$

Then we have the following Corollary

Corollary 11

Let the hypotheses i)-v) are satisfied. Then a nonlinear equation of convolution type

$$x(t) = \int_0^t k_1(t,s)f\left(s, \int_0^s k_2(s-\theta)x(\varphi(\theta))d\theta\right)ds,$$

$$t \in R_+$$

(12)

has at least one integrable solution $x \in L_1$.

In the next subsection, we prove an existence theorem for integral equation of fractional order as a special form of Equation (12).

Initial Value Problems of Fractional Order

As a special case of Equation (14), we consider

$$x(t) = \int_0^t k_1(t,s) f\left(s, \int_0^s \frac{(s-\theta)^{-\beta}}{\Gamma(1-\beta)} x(\theta) d\theta\right) ds, t \in R_+$$

(13)

where

$$k_2(s-\theta) = \frac{(s-\theta)^{-\beta}}{\Gamma(1-\beta)}, \beta \in (0,1]$$

And $\varphi(\theta) = \theta$. Equation (13) is an integral equation of fractional order that can be written in the form

$$x(t) = \int_0^t k_1(t,s) f\left(s, I^{1-\beta} x(s)\right) ds, t \in R_+$$

(14)

Obviously, Equation (14) has at least one integrable solution $x \in L_1$.

Definition 12

By a solution of the initial value problem (1) we mean an absolutely continuous function x satisfies the initial value problem (1).

Theorem 13

Let $b\|K_2\|/\Gamma(2-\beta)<1$ and $\beta \in (0,1.]$.

If assumptions i)-iii) and v) are satisfied, then the initial value problem (1) has at least one solution $y \in L_1$.

Proof

Let x be a solution of the integral Equation (14). Putting

$$y(t) = \int_0^t x(\tau)d\tau.$$

Since x is integrable, then

$$Dy(t) = D\int_0^t x(\tau)d\tau \text{ a.e.}$$

Where $D = \dfrac{d}{dt}$. Moreover, the integral $\int_0^t x(\tau)d\tau$ of integrable function x is absolutely continuous then

$$Dy(t) = DI^1 x(t)$$

Then we have,

$$Dy(t) = x(t) \text{ a.e.}$$

Furthermore, we obtain

$$Dy(t) = x(t)$$
$$I^{1-\beta} Dy(t) = I^{1-\beta} x(t)$$
$$D^{\beta} y(t) = I^{1-\beta} x(t)$$

Consequently, Equation (14) gives

$$\frac{d}{dt} y(t) = Dy(t) = \int_0^t k_1(t,s) f(s, D^{\beta} y(s)) ds,$$

Since x is integrable and absolutely continuous, then

$$Dy(\tau) = x(\tau)$$
$$I^1 Dy(\tau) = I^1 x(\tau)$$
$$\int_0^t \frac{d}{dt} y(\tau) d\tau = \int_0^t x(\tau) d\tau$$

Clearly $y(0) = y_0$ Hence we deduce that y is an absolutely continuous function satisfies the initial value problem (1). Hence the proof is complete.

CONCLUSIONS

The existence theorem of functional integrable equation in the space of Lebesgue integrable functions on unbounded interval $L_1[0, \infty)$ is presented and proved. As an application of this theorem, we investigated the existence of solution of the suggested initial value problems of fractional order.

REFERENCES

1. J. Banas and Z. Knap, "Integrable Solutions of a Functional-Integral Equation," Revista Matemática de la Universidad Complutense de Madrid, Vol. 2, No. 1, 1989, pp. 31-38.

2. J. Banas and A. Chlebowicz, "On Existence of Integrable Solutions of a Functional Integral Equation under Carathéodory Conditions," Nonlinear Analysis, Vol. 70, No. 9, 2009, pp. 3172-3179.

3. G. Emmanuele, "Integrable Solutions of a Functional Integral Equation," Journal of Integral Equations and Applications, Vol. 4, No. 1, 1992, pp. 89-94. doi:10.1216/jiea/1181075668

4. P. P. Zabrejko, A. I. Koshelev, M. A. Krasnosel'skii, S. G. Mikhlin, L. S. Rakovshchik and V. J. Stecenko, "Integral Equations," Noordhoff, Leyden, 1975.

5. A. M. A. El-Sayed, "Nonlinear Functional Differential Equations of Arbitrary Orders," Nonlinear Analysis, Vol. 33, No. 2, 1998, pp. 181-186. doi:10.1016/S0362-546X(97)00525-7

6. A. M. A. El-Sayed, N. Sherif and I. A. Ibrahim, "On a Mixed Type Integral Equation and Fractional Order Functional Differential Equations," Commentationes Mathematicae. Prace Matematyczne, Vol. 45, No. 2, 2005, pp. 237-247.

7. I. A. Ibrahim, T. S. Amer and Y. M. Abo Essa, "Integrable Solutions of Initial Value Problems of Fractional Order," Far East Journal of Mathematical Sciences, Vol. 62, No. 1, 2012, pp. 97-123.

8. J. Appel, "Implicit Functions, Nonlinear Integral Equations and the Measure of Noncompactness of the Superposition Operator," Journal of Mathematical Analysis and Applications, Vol. 83, No. 1, 1981, pp. 251-263. doi:10.1016/0022-247X(81)90261-4

9. M. A. Krasnosel'skii, P. P. Zabrejko, J. I. Pustyl'nik and P. J. Sobolevskii, "Integral Operators in Spaces of Summable Functions," Noordhoff, Leyden, 1976.

10. R. Pluciennik, "On Some Properties of the Superposition Operator in Generalized Orlicz Spaces of Vector-Valued Functions," Commentationes Mathematicae. Prace Matematyczne, Vol. 25, No. 2, 1985, pp. 321-337.

11. K. Carathéodory, "Vorlesungen Über Reele Funktionen," De Gruyter, Leipzig, 1918.

12. J. Appel and P. P. Zabrejko, "Nonlinear Superposition Operators," In: Cambridge Tracts in Mathematics, Vol. 95, Cambridge University Press, Cambridge, 1990.

13. G. Scorza Dragoni, "Un Teorema Sulle Funzioni Continue Rispetto ad une e Misarubili Rispetto ad Un'altra Variable," Rendiconti del Seminario Matematico della Università di Padova, 1948, pp. 102-106.

14. N. Dunford and J. T. Schwartz, "Linear Operators," Int. Publ., Leyden, 1963.

15. J. Banas and W. G. El-Sayed, "Measures of Noncompactness and Solvability of an Integral Equation in the Class of Functions of Locally Bounded Variaton," Journal of Mathematical Analysis and Applications, Vol. 167, No. 1, 1992, pp. 133-151.doi:10.1016/0022-247X(92)90241-5

16. J. Banas and J. Rivero, "On Measures of Weak Noncompactness," Annali di Matematica Pura ed Applicata, Vol. 151, No. 1, 1988, pp. 213-224. doi:10.1007/BF01762795

17. J. Banas and Z. Knap, "Measures of Weak Noncompactness and Nonlinear Integral Equations of Convolution Type," Journal of Mathematical Analysis and Applications, Vol. 146, No. 2, 1990, pp. 353-362. doi:10.1016/0022-247X(90)90307-2

18. J. Dieudonné, "Sur les Espaces de Köthe," Journal d'Analyse Mathématique, Vol. 1, No. 1, 1951, pp. 81-115. doi:10.1007/BF02790084

19. A. M. A. El-Sayed, W. G. El-Sayed and O. L. Moustafa, "On Some Fractional Functional Equations," Pure Mathematics and Applications, Vol. 6, No. 4, 1995, pp. 321- 332.

20. S. G. Samko, A. A. Kilbas and O. I. Marichev, "Fractional Integrals and Derivatives Theory and Applications," Gordon and Breach Science Publishers, Amsterdam, 1993.

CITATION

I. Ibrahim, T. Amer and Y. Aboessa, "Existence Theorem for a Nonlinear Functional Integral Equation and an Initial Value Problem of Fractional Order in $L_1(R_+)$," Applied Mathematics, Vol. 4 No. 2, 2013, pp. 402-409. doi:10.4236/am.2013.42060.

Applications of Multivalent Functions Associated with Generalized Fractional Integral Operator

Jae Ho Choi
Department of Mathematics Education,
Daegu National University of Education,
Daegu, South Korea

ABSTRACT

By using a method based upon the Briot-Bouquet differential subordination, we investigate some subordination properties of the generalized fractional integral operator $\mathcal{J}_{0,z}^{\lambda,\mu,\nu}$ which was defined by Owa, Saigo and Srivastava [1]. Some interesting further consequences are also considered.

INTRODUCTION

Let $A_n(p)$ denote the class of functions $f(z)$ of the form

$$f(z) = z^p + \sum_{r}^{\infty} a_{p+k} z^{p+k}, \left(p, n \in \mathbb{N} := \{1, 2, 3, \cdots\}\right), \tag{1.1}$$

which are analytic in the open unit disk $\mathbb{U} = \{z : z \in \mathbb{C} \text{ and } |z| < 1\}$ Also let f and g be analytic in $\mathbb{U}$ with f(0)=g(0). Then we say that f is subordinate to g in $\mathbb{U}$, written $f \prec g$ or $f(z) \prec g(z)$, if there exists the Schwarz function w, analytic in $\mathbb{U}$ such that w(0)=0, $|w(z)| < 1$ and $f(z) = g(w(z))(z \in \mathbb{U})$. We also observe that

$$f(z) \prec g(z) \text{ in } \mathbb{U}$$

if and only if

$$f(0) = g(0) \text{ and } f(\mathbb{U}) \subset g(\mathbb{U})$$

whenever g is univalent in U.

Let a, b and c be complex numbers with $c \neq 0, -1, -2, \cdots$. Then the Gaussian/classical hypergeometric function $_2F_1(a, b; c; z)$ is defined by

$$_2F_1(a, b; c; z) = \sum_{k=0}^{\infty} \frac{(a)_k (b)_k}{(c)_k} \frac{z^k}{k!},$$

(1.2)

where $(\eta)_k$ is the Pochhammer symbol defined, in terms of the Gamma function, by

$$(\eta)_k = \frac{\Gamma(\eta + k)}{\Gamma(\eta)} = \begin{cases} 1, & (k = 0) \\ \eta(\eta + 1) \cdots (\eta + k - 1), & (k \in \mathbb{N}). \end{cases}$$

(1.3)

The hypergeometric function $_2F_1(a, b; c; z)$ is analytic in U and if a or b is a negative integer, then it reduces to a polynomial.

For each A and B such that $-1 \leq B < A \leq 1$, let us define the function

$$h(A, B; z) = \frac{1 + Az}{1 + Bz}, (z \in \mathbb{U}).$$

(1.4)

It is well known that h(A,B;z), for $-1 \leq B \leq 1$, is the conformal map of the unit disk onto the disk symmetrical respect to the real axis having the center $(1-AB)/(1-B^2)$ and the radius $(A-B)/(1-B^2)$. The boundary circle cuts the real axis at the points $(1-B)/(1-B)$ and $(1+B)/(1+B)$.

Many essentially equivalent definitions of fractional calculus have been given in the literature (cf., e.g. [2, 3]). We state here the following definition due to Saigo [4] (see also [1, 5]).

Definition 1:

For $\lambda > 0, \lambda > 0, \mu, \nu \in \mathbb{R}$, the fractional integral operator $\mathcal{J}_{0,z}^{\lambda,\mu,\nu}$ is defined by

$$\mathcal{I}_{0,z}^{\lambda,\mu,\nu} f(z)$$

$$= \frac{z^{-\lambda-\mu}}{\Gamma(\lambda)} \int_0^z (z-\varsigma)_2^{\lambda-1} F_1\left(\lambda+\mu,-\nu;\lambda;1-\frac{\varsigma}{z}\right) f(\varsigma) d\varsigma,$$

$$(1.5)$$

where $_2F_1$ is the Gaussian hypergeometric function defined by (1.2) and $f(z)$ is taken to be an analytic function in a simply-connected region of the z-plane containing the origin with the order

$$f(z) = \mathcal{O}\left(|z|^\epsilon\right)(z \to 0)$$

For $\epsilon > \max\{0, \mu - \nu\} - 1$, and the multiplicity of $(z-\varsigma)^{\lambda-1}$ is removed by requiring that $\log(z-\varsigma)$ to be real when $z-\varsigma > 0$.

The definition (1.5) is an interesting extension of both the Riemann-Liouville and Erdélyi-Kober fractional operators in terms of Gauss's hypergeometric functions.

With the aid of the above definition, Owa, Saigo and Srivastava [1] defined a modification of the fractional integral operator $\mathcal{J}_{0,z}^{\lambda,\mu,\nu}$ by

$$\mathcal{J}_{0,z}^{\lambda,\mu,\nu} f(z)$$
$$= \frac{\Gamma(p+1-\mu)\Gamma(\lambda+p+1+\nu)}{\Gamma(p+1)\Gamma(p+1-\mu+\nu)} z^{\mu} \mathcal{I}_{0,z}^{\lambda,\mu,\nu} f(z) \tag{1.6}$$

for $f(z) \in A_n(p)$ and $\mu - \nu - p < 1$. Then it is observed that $\mathcal{J}_{0,z}^{\lambda,\mu,\nu}$ also maps $A_n(p)$ onto itself as follows:

$$\mathcal{J}_{0,z}^{\lambda,\mu,\nu} f(z)$$
$$= z^p + \sum_{k=n}^{\infty} \frac{(p+1)_k (p+1-\mu+\nu)_k}{(p+1-\mu)_k (\lambda+p+1+\nu)_k} a_{p+k} z^{p+k},$$
$$(\lambda > 1; \mu - \nu - p < 1; f \in A_n(p)). \tag{1.7}$$

We note that $\mathcal{J}_{0,z}^{\alpha,0,\beta-1} f(z) = \mathcal{O}_{\beta}^{\alpha} f(z), (\alpha \geq 0; \beta > -1)$, where the operator $\mathcal{O}_{\beta}^{\alpha}$ was introduced and studied by Jung, Kim and Srivastava [6] (see also [7]).

It is easily verified from (1.7) that

$$z \left(\mathcal{J}_{0,z}^{\lambda,\mu,\nu} f(z) \right)'$$
$$= (\lambda + \nu + p) \mathcal{J}_{0,z}^{\lambda-1,\mu,\nu} f(z) - (\lambda + \nu) \mathcal{J}_{0,z}^{\lambda,\mu,\nu} f(z). \tag{1.8}$$

The identity (1.8) plays an important and significant role in obtaining our results.

Recently, by using the general theory of differential subordination, several authors (see, e.g. [7-9]) considered some interesting properties of multivalent functions associated with various integral operators. In this manuscript, we shall derive some subordination properties of the fractional integral operator $\mathcal{J}_{0,z}^{\lambda,\mu,\nu}$ by using the technique of differential subordination.

MAIN RESULTS

In order to establish our results, we shall need the following lemma due to Miller and Mocanu [10].

Lemma 1:

Let h(t) be analytic and convex univalent in $\mathbb{U}$ with h(0)=1, and let $g(z) = 1 + b_n z^n + b_n z^{n+1} + \cdots$ be analytic in $\mathbb{U}$. If

$$g(z) + \frac{1}{c} z g'(z) \prec h(z), \qquad (2.1)$$

Then for $c \neq 0$ and $\mathrm{Re}\, c \geq 0,,$

$$g(z) \prec \frac{c}{n} z^{-c/n} \int_0^z t^{c/n-1} h(t) \, dt. \qquad (2.2)$$

We begin by proving the following theorem.

Theorem 1:

Let, $-1 \leq B < A \leq 1, \lambda > 1, \lambda + v > -p, \mu - v - p < 1, \mu - 1 < p$ and $0 < \alpha < 1$, and let

$$f(z) = z^p + \sum_{k=n}^{\infty} a_{p+k} z^{p+k} \in A_n(p) \text{ Suppose that}$$

$$\sum_{k=n}^{\infty} c_k \left| a_{p+k} \right| \leq 1, \qquad (2.3)$$

where

$$c_k = \frac{1-B}{A-B} \frac{\left[\lambda + p + v + k(1-\alpha)\right](p+1)_k \, (p+1-\mu+v)_k}{(\lambda + p + v)(p+1-\mu)_k \, (\lambda + p + 1 + v)_k} \tag{2.4}$$

and $(\eta)_k$ is given by (1.3).

1) If $-1 \leq B < 0$, then

$$(1-\alpha)\frac{\mathcal{J}_{0,z}^{\lambda-1,\mu,v} f(z)}{z^p} + \alpha \frac{\mathcal{J}_{0,z}^{\lambda,\mu,v} f(z)}{z^p} \prec h(A,B;z). \tag{2.5}$$

2) If $-1 \leq B < 0$ and $\gamma \geq 1$, then

$$\mathrm{Re}\left\{ \left(\frac{\mathcal{J}_{0,z}^{\lambda,\mu,v} f(z)}{z^p} \right)^{1/\gamma} \right\}$$

$$> \left\{ \frac{\lambda+v+p}{n(1-\alpha)} \int_0^1 u^{\frac{\lambda+v+p}{n(1-\alpha)}-1} \left(\frac{1-Au}{1-Bu} \right) du \right\}^{1/\gamma} , (z \in U). \tag{2.6}$$

The result is sharp.

Proof

1. If we set

$$L = (1-\alpha)\frac{\mathcal{J}_{0,z}^{\lambda-1,\mu,v} f(z)}{z^p} + \alpha \frac{\mathcal{J}_{0,z}^{\lambda,\mu,v} f(z)}{z^p},$$

then, from (1.7) we see that

$$L = 1 + \sum_{k=n}^{\infty} \frac{\left[\lambda + p + v + k(1-\alpha)\right](p+1)_k \left(p+1-\mu+v\right)_k}{(\lambda+p+v)(p+1-\mu)_k (\lambda+p+1+v)_k} a_{p+k} z^k.$$

(2.7)

For $-1 \le B < 0$ and $z \in \mathbb{U}$, it follows from (2.3) that

$$\left|\frac{L-1}{A-BL}\right| = \left|\frac{\sum\limits_{k=n}^{\infty} \frac{\left[\lambda + p + v + k(1-\alpha)\right](p+1)_k \left(p+1-\mu+v\right)_k}{(\lambda+p+v)(p+1-\mu)_k (\lambda+p+1+v)_k} a_{p+k} z^k}{A - B - B\sum\limits_{k=n}^{\infty} \frac{\left[\lambda + p + v + k(1-\alpha)\right](p+1)_k \left(p+1-\mu+v\right)_k}{(\lambda+p+v)(p+1-\mu)_k (\lambda+p+1+v)_k} a_{p+k} z^k}\right| \le \frac{\sum\limits_{k=n}^{\infty} c_k |a_{p+k}|}{1 - B + B\sum\limits_{k=n}^{\infty} c_k |a_{p+k}|} \le 1.$$

(2.8)

which implies that

$$(1-\alpha)\frac{\mathcal{J}_{0,z}^{\lambda-1,\mu,v} f(z)}{z^p} + \alpha \frac{\mathcal{J}_{0,z}^{\lambda,\mu,v} f(z)}{z^p} \prec h(A,B;z).$$

2. Let

$$g(z) = \frac{\mathcal{J}_{0,z}^{\lambda,\mu,v} f(z)}{z^p}, \left(f \in \mathcal{A}_n(p)\right).$$

(2.9)

Then the function $g(z) = 1 + b_n z^n + b_{n+1} z^{n+1} + \cdots$ is analytic in $\mathbb{U}$. Using (1.8) and (2.9), we have

$$\frac{\mathcal{J}_{0,z}^{\lambda-1,\mu,v} f(z)}{z^p} = g(z) + \frac{1}{\lambda+v+p} zg'(z).$$

(2.10)

From (2.5), (2.9) and (2.10) we obtain

$$(1-\alpha)\frac{\mathcal{J}_{0,z}^{\lambda-1,\mu,\nu}f(z)}{z^p}+\alpha\frac{\mathcal{J}_{0,z}^{\lambda,\mu,\nu}f(z)}{z^p}$$

$$=g(z)+\frac{1-\alpha}{\lambda+\nu+p}zg'(z)\prec h(A,B;z).$$

Thus, by applying Lemma 1, we observe that

$$g(z)\prec\frac{\lambda+\nu+p}{n(1-\alpha)}z^{-\frac{\lambda+\nu+p}{n(1-\alpha)}}\int_0^z t^{\frac{\lambda+\nu+p}{n(1-\alpha)}-1}\left(\frac{1+At}{1+Bt}\right)dt$$

or

$$\frac{\mathcal{J}_{0,z}^{\lambda,\mu,\nu}f(z)}{z^p}=\frac{\lambda+\nu+p}{n(1-\alpha)}\int_0^1 u^{\frac{\lambda+\nu+p}{n(1-\alpha)}-1}\left(\frac{1+Auw(z)}{1+Buw(z)}\right)du,$$

$$\tag{2.11}$$

where w(z) is analytic in $\mathbb{U}$ with w(0)=0 and $|w(z)|<1(z\in\mathbb{U})$. In view of $-1\le B<A\le 1$ and $\lambda+\nu>-p$, we conclude from (2.11) that

$$\mathrm{Re}\left\{\frac{\mathcal{J}_{0,z}^{\lambda,\mu,\nu}f(z)}{z^p}\right\}>\frac{\lambda+\nu+p}{n(1-\alpha)}\int_0^1 u^{\frac{\lambda+\nu+p}{n(1-\alpha)}-1}\left(\frac{1-Au}{1-Bu}\right)du,$$

$$(z\in\mathbb{U}).$$

$$\tag{2.12}$$

Since $\mathrm{Re}(w^{1/\gamma})\ge(\mathrm{Re}w)^{1/\gamma}$ for $\mathrm{Re}w>0$ and $\gamma\ge 1$, from (2.12) we see that the inequality (2.6) holds.

To prove sharpness, we take $f(z) \in A_n(p)$ defined by

$$\frac{\mathcal{J}_{0,z}^{\lambda,\mu,v} f(z)}{z^p} = \frac{\lambda+v+p}{n(1-\alpha)} \int_0^1 u^{\frac{\lambda+v+p}{n(1-\alpha)}-1} \left(\frac{1+Auz^n}{1+Buz^n} \right) du.$$

For this function we find that

$$(1-\alpha) \frac{\mathcal{J}_{0,z}^{\lambda-1,\mu,v} f(z)}{z^p} + \alpha \frac{\mathcal{J}_{0,z}^{\lambda,\mu,v} f(z)}{z^p} = \frac{1+Az^n}{1+Bz^n}$$

and

$$\frac{\mathcal{J}_{0,z}^{\lambda,\mu,v} f(z)}{z^p} \to \frac{\lambda+v+p}{n(1-\alpha)} \int_0^1 u^{\frac{\lambda+v+p}{n(1-\alpha)}-1} \frac{1-Au}{1-Bu} du \text{ as } z \to e^{i\pi/n}.$$

Hence the proof of Theorem 1 is evidently completed.

Theorem 2:

Let, $-1 \leq B < A \leq 1, \lambda > 1, \lambda+v > -p, \mu-v-p < 1, \mu-1 < p$ and $0 < \alpha < 1$.

Suppose that $f(z) = z^p + \sum_{k=n}^{\infty} a_{p+k} z^{p+k} \in A_n(p)$, $S_1(z) = Z^p$ and $S_m(z) =$

$Z^p + \sum_{k=n}^{n+m-2} a_{p+k} z^{p+k}$ $(m \geq 2)$. If the sequence $\{C_k\}$ is nondecreasing with

$$c_k \geq \frac{(1-B)\left[\lambda+p+v+k(1-\alpha)\right]}{(A-B)(\lambda+p+v)} (k \geq n),$$

$$(2.13)$$

where C_k is given by (2.4) and satisfies the condition (2.4), then

$$\text{Re}\left\{\frac{\mathcal{J}_{0,z}^{\lambda,\mu,\nu}f(z)}{s_m(z)}\right\} > 0$$

(2.14)

and

$$\text{Re}\left\{\frac{s_m(z)}{\mathcal{J}_{0,z}^{\lambda,\mu,\nu}f(z)}\right\} > 0.$$

(2.15)

Each of the bounds in (2.14) and (2.15) is best possible for $m \in \mathbb{N}$.

Proof:

We prove the bound in (2.14). The bound in (2.15) is immediately obtained from (2.14) and will be omitted. Let

$$h(z) = \frac{\mathcal{J}_{0,z}^{\lambda,\mu,\nu}f(z)}{s_m(z)} \left(f \in \mathcal{A}_n(p); z \in \mathbb{U} \right).$$

Then, from (1.7) we observe that

$$h(z) = 1 + \frac{\displaystyle\sum_{k=n}^{n+m-2}(\delta_k-1)a_{p+k}z^k + \sum_{k=n+m-1}^{\infty}\delta_k a_{p+k}z^k}{1 + \displaystyle\sum_{k=n}^{n+m-2}a_{p+k}z^k},$$

where, for convenience,

$$\delta_k = \frac{(p+1)_k (p+1-\mu+v)_k}{(p+1-\mu)_k (\lambda+p+1+v)_k}.$$

It is easily seen from (2.4) and (2.13) that $C_k > 1$ and

$$\delta_k = \frac{(A-B)(\lambda+p+v)}{(1-B)\left[\lambda+p+v+k(1-\alpha)\right]} C_k \geq 1. \qquad (2.16)$$

Hence, by applying (2.3) and (2.16), we have

$$\left|\frac{h(z)-1}{h(z)+1}\right| = \frac{\left|\sum_{k=n}^{n+m-2} (\delta_k - 1) a_{p+k} z^k + \sum_{k=n+m-1}^{\infty} \delta_k a_{p+k} z^k\right|}{\left|2 + \sum_{k=n}^{n+m-2} (\delta_k + 1) a_{p+k} z^k + \sum_{k=n+m-1}^{\infty} \delta_k a_{p+k} z^k\right|}$$

$$\leq \frac{\sum_{k=n}^{n+m-2} (\delta_k - 1)\left|a_{p+k}\right| + \sum_{k=n+m-1}^{\infty} \delta_k \left|a_{p+k}\right|}{2 - \sum_{k=n}^{n+m-2} (\delta_k + 1)\left|a_{p+k}\right| - \sum_{k=n+m-1}^{\infty} \delta_k \left|a_{p+k}\right|} \leq 1 (z \in \mathbb{U})$$

which readily yields the inequality (2.14).

If we take $f(z) = Z^p - Z^{p+n+m-1}$, then

$$\frac{f(z)}{S_m(z)} = 1 - z^{n+m-1} \rightarrow 0 \text{ as } z \rightarrow 1^-.$$

This show that the bound in (2.14) is best possible for each m, which proves Theorem 2.

Finally, we consider the generalized Bernardi-LiveraLivingston integral operator $\mathcal{L} \, L_\sigma (\sigma > -p)$ defined by (cf. [11-13])

$$L_\sigma (f)(z) := \frac{\sigma + p}{z^\sigma} \int_0^z t^{\sigma-1} f(t) \, dt \left(f \in A_n(p); \sigma > -p \right). \tag{2.17}$$

Theorem 3:

Let, $-1 \geq B < A \geq 1, \sigma > -p, \lambda > 1 \lambda + v > -p, \mu - v - p < 1, \quad \mu - 1 < p$ and $0 < \alpha < 1$ and let $f(z) = z^p + \sum\limits_{k=n}^{\infty} a_{p+k} z^{p+k} \in A_n(p)$. Suppose that

$$\sum_{k=n}^{\infty} d_k \left| a_{p+k} \right| \leq 1,$$

$$\tag{2.18}$$

where

$$d_k = \frac{1-B}{A-B} \frac{\left[\sigma + p + k(1-\alpha)\right](p+1)_k (p+1-\mu+v)_k}{(\sigma+p+k)(p+1-\mu)_k (\lambda+p+1+v)_k}$$

and $(\eta)_k$ is given by (1.3).

1) If $-1 \leq B < 0$, then

$$(1-\alpha)\frac{\mathcal{J}_{0,z}^{\lambda,\mu,v} f(z)}{z^p} + \alpha\frac{\mathcal{J}_{0,z}^{\lambda,\mu,v} L_\sigma (f)(z)}{z^p} \prec h(A,B;z). \tag{2.19}$$

2) If $-1, \leq B < 0$ and $\gamma \geq 1$, then

$$\text{Re}\left\{\left(\frac{\mathcal{J}_{0,z}^{\lambda,\mu,v}\mathcal{L}_\sigma(f)(z)}{z^p}\right)^{1/\gamma}\right\}$$

$$> \left\{\frac{\sigma+p}{n(1-\alpha)}\int_0^1 u^{\frac{\sigma+p}{n(1-\alpha)}-1}\left(\frac{1-Au}{1-Bu}\right)du\right\}^{1/\gamma} \quad (z \in U).$$

$$(2.20)$$

The result is sharp.

Proof:

1. If we put

$$M = (1-\alpha)\frac{\mathcal{J}_{0,z}^{\lambda,\mu,v}f(z)}{z^p} + \alpha\frac{\mathcal{J}_{0,z}^{\lambda,\mu,v}\mathcal{L}_\sigma(f)(z)}{z^p},$$

then, from (1.7) and (2.17) we have

$$M = 1 + \sum_{k=n}^{\infty}\frac{\left[\sigma+p+k(1-\alpha)\right](p+1)_k(p+1-\mu+v)_k}{(\sigma+p+k)(p+1-\mu)_k(\lambda+p+1+v)_k}$$
$$\cdot a_{p+k}z^k.$$

Therefore, by using same techniques as in the proof of Theorem 1 1), we obtain the desired result.

2. From (2.17) we have

$$(\sigma+p)\mathcal{J}_{0,z}^{\lambda,\mu,v}f(z)$$

$$= \sigma\mathcal{J}_{0,z}^{\lambda,\mu,v}\mathcal{L}_\sigma(f)(z) + z\left(\mathcal{J}_{0,z}^{\lambda,\mu,v}\mathcal{L}_\sigma(f)(z)\right)'.$$

$$(2.21)$$

Let

$$g(z) = \frac{\mathcal{J}_{0,z}^{\lambda,\mu,\nu} \mathcal{L}_\sigma (f)(z)}{z^p} \quad (z \in \mathbb{U}).$$

(2.22)

Then, by virtue of (2.21), (2.22) and (2.19), we observe that

$$(1-\gamma)\frac{\mathcal{J}_{0,z}^{\lambda,\mu,\nu} f(z)}{z^p} + \gamma\frac{\mathcal{J}_{0,z}^{\lambda,\mu,\nu} \mathcal{L}_\sigma (f)(z)}{z^p}$$

$$= g(z) + \frac{1-\gamma}{\sigma+p} zg'(z) \prec h(A,B;z).$$

Hence, by applying the same argument as in the proof of Theorem 1 2), we obtain (2.20), which evidently proves Theorem 3.

ACKNOWLEDGMENTS

This work was supported by Daegu National University of Education Research grant in 2011.

REFERENCES

1. S. Owa, M. Saigo and H. M. Srivastava, "Some Characterization Theorems for Starlike and Convex Functions Involving a Certain Fractional Integral Operator," Journal of Mathematical Analysis and Applications, Vol. 140, No. 2, 1989, pp. 419-426.doi:10.1016/0022-247X(89)90075-9

2. S. G. Samko, A. A. Kilbas and O. I. Marichev, "Fractional Integral and Derivatives, Theory and Applications," Gordon and Breach, New York, Philadelphia, London, Paris, Montreux, Toronto, Melbourne, 1993.

3. H. M. Srivastava and R. G. Buschman, "Theory and Applications of Convolution Integral Equations," Kluwer Academic Publishers, Dordrecht, Boston, London, 1992.

4. M. Saigo, "A Remark on Integral Operators Involving the Gauss Hypergeometric Functions," Mathematical Reports, Kyushu University, Vol. 11, No. 2, 1977-1978, pp. 135- 143.

5. J. H. Choi, "Note on Differential Subordination Associated with Fractional Integral Operator," Far East Journal of Mathematical Sciences, Vol. 26, No. 2, 2007, pp. 499- 511.

6. B. Jung, Y. C. Kim and H. M. srivastava, "The Hardy Space of Analytic Functions Associated with Certain OneParameter Families of Integral Operators," Journal of Mathematical Analysis and Applications, Vol. 176, No. 1, 1993, pp. 138-147. doi:10.1006/jmaa.1993.1204

7. J.-L. Liu, "Notes on Jung-Kim-Srivastava Integral Operator," Journal of Mathematical Analysis and Applications, Vol. 294, No. 1, 2004, pp. 96-103.doi:10.1016/j.jmaa.2004.01.040

8. R. M. EL-Ashwash and M. K. Aouf, "Some Subclasses of Multivalent Functions Involving the Extended Fractional Differintegral Operator," Journal of Mathematical Inequalities, Vol. 4, No. 1, 2010, pp. 77-93.

9. J. Patel, A. K. Mishra and H. M. Srivastava, "Classes of Multinalent Analytic Functions Involving the DziokSrivastava Operator," Computers and Mathematics with Applications, Vol. 54, No. 5, 2007, pp. 599-616. doi:10.1016/j.camwa.2006.08.041

10. S. S. Miller and P. T. Mocanu, "Differential Subordinations and Univalent Functions," Michigan Mathematical Journal, Vol. 28, No. 2, 1981, pp. 157-172. doi:10.1307/mmj/1029002507

11. S. D. Bernardi, "Convex and Starlike Univalent Functions," Transactions of the American Mathematical Society, Vol. 135, 1969, pp. 429-446. doi:10.1090/S0002-9947-1969-0232920-2

12. R. J. Libera, "Some Classes of Regular Univalent Functions," Proceedings of the American Mathematical Society, Vol. 16, No. 4, 1965, pp. 755-758. doi:10.1090/S0002-9939-1965-0178131-2

13. H. M. Srivastava and S. Owa, Eds., "Current Topics in Analytic Function Theory," World Scientific Publishing Company, Singapore, New Jersey, London, Hong Kong, 1992.doi:10.1142/1628

CITATION

J. Choi, "Applications of Multivalent Functions Associated with Generalized Fractional Integral Operator," Advances in Pure Mathematics, Vol. 3 No. 1, 2013, pp. 1-5. doi: 10.4236/apm.2013.31001.

Mapping Properties of Generalized Robertson Functions under Certain Integral Operators

Muhammad Arif, Wasim Ul-Haq, and Muhammad Ismail
Department of Mathematics, Abdul Wali
Khan University, Mardan, Pakistan

6

ABSTRACT

In the present article, certain classes of generalized p-valent Robertson functions are considered. Mapping properties of these classes are investigated under certain p-valent integral operators introduced by Frasin recently.

INTRODUCTION

Let A(p) be the class of functions f(z) of the form

$$f(z) = z^p + \sum_{j=p+1}^{\infty} a_j z^j \quad (p \in \mathbb{N} = \{1, 2, \cdots\}),$$

which are analytic in the open unit disc $E = \{z : |z| < 1\}$. We write A(1)=A.

A function $f(z) \in A$ is said to be spiral-like if there exists a real number

$$\lambda \left(|\lambda| < \frac{\pi}{2} \right)$$

such that

$$\text{Re } e^{i\lambda} \frac{zf'(z)}{f(z)} > 0 \ (z \in E).$$

The class of all spiral-like functions was introduced by L. Spacek [1] in 1933 and we denote it by S_λ^*. Later in 1969, Robertson [2] considered the class C_λ of analytic functions in E for which $zf'(z) \in S_\lambda^*$.

Let $P_k^\lambda(p,\rho)$ be the class of functions p(z) analytic in E with p(0)=1 and

$$\int_0^{2\pi} \left| \frac{\text{Re } e^{i\lambda} p(z) - \rho \cos \lambda}{p - \rho} \right| d\theta \le k\pi \cos \lambda, \ \left(z = re^{i\theta} \right),$$

where $k \ge 2$, $0 \le \rho < 1$ and λ is real with $|\lambda| < \dfrac{\pi}{2}$.

For $\lambda = 0$, p=1, this class was introduced in [3] and for $\rho = 0$, see [4]. For k=2, $\lambda = 0$ and $\rho = 0$, the class $P_k^\lambda(p,\rho)$ reduces to the class P of functions P(Z) analytic in E with p(0)=1 and whose real part is positive.

We define the following classes

$$R_k^\lambda(p,\rho) = \left\{ f(z) : f(z) \in A(p) \text{ and } \frac{zf'(z)}{f(z)} \in P_k^\lambda(p,\rho), \ 0 \le \rho < 1 \right\},$$

$$V_k^\lambda(p,\rho) = \left\{ f(z) : f(z) \in A(p) \text{ and } \frac{\left(zf'(z) \right)'}{f'(z)} \in P_k^\lambda(p,\rho), \ 0 \le \rho < 1 \right\}.$$

For $\lambda = 0$, $\rho = 0$ and p=1, we obtain the well-known classes R_k and V_k of analytic functions with bounded radius and bounded boundary rotations studied by Tammi [5] and Paatero [6] respectively. For details see [7-12]. Also it can easily be seen that $R_{\frac{\lambda}{2}}(0)=S_{\lambda}^{*}$ and $V_{\frac{\lambda}{2}}(0)=C_{\lambda}$

Let us consider the integral operators

$$F_p(z)=\int_0^z pt^{p-1}\left(\frac{f_1(t)}{t^p}\right)^{\alpha_1}\cdots\left(\frac{f_n(t)}{t^p}\right)^{\alpha_n} dt \tag{1.1}$$

and

$$G_p(z)=\int_0^z pt^{p-1}\left[\frac{f_1'(t)}{pt^{p-1}}\right]^{\alpha_1}\cdots\left[\frac{f_n'(t)}{pt^{p-1}}\right]^{\alpha_n} dt, \tag{1.2}$$

where $f_i(z)\in A(p)$ and $\alpha_i>0$ for all $i\in\{1,2,\ldots\ldots,n\}$.

These operators, given by (1.1) and (1.2), are defined by Frasin [13]. If we take $p=1$, we obtain the integral operators $F_1(z)=F_n(z)$ and $G_1(z)=F_{\alpha_1\ldots\alpha_n}(z)$ introduced and studied by Breaz and Breaz [14] and Breaz et al. [15], for details see also [16-20]. Also for $p=n=1$, $\alpha_1=\alpha\in[0,1]$ in (1.1), we obtain the integral operator studied in [21] given as

$$\int_0^z\left(\frac{f(t)}{t}\right)^{\alpha} dt,$$

and for $p=n=1$, $\alpha_1=\delta\in\mathbb{C}$, $|\delta|\le\frac{1}{4}$ in (1.2), we obtain the integral operator

$$\int_0^z(f'(t))^{\delta} dt,$$

discussed in [22,23].

In this paper, we investigate some propeties of the above integral operators $F_p(z)$ and $G_p(z)$ for the classes $V_k^\lambda(p,\rho)$ and $R_k^\lambda(p,\rho)$ respectively.

MAIN RESULT

Theorem 2.1:

Let $f_i(z) \in R_k^\lambda(p,\rho)$ for $1 \le i \le n$ with $0 \le \rho < 1$. Also let λ is real with

$$|\lambda| < \frac{\pi}{2}; \alpha_i > 0,$$

$1 \le i \le n$. If

$$0 \le (\rho - p) \sum_{i=1}^{n} \alpha_i + p < 1,$$

then $F_p(z) \in V_k^\lambda(p,\lambda_1)$ with

$$\lambda_1 = (\rho - p) \sum_{i=1}^{n} \alpha_i + p. \tag{2.1}$$

Proof: From (1.1), we have

$$\frac{zF_p''(z)}{F_p'(z)} = (p-1) + \sum_{i=1}^{n} \alpha_i \left(\frac{zf_i'(z)}{f_i(z)} - p \right), \tag{2.2}$$

or, equivalently

$$e^{i\lambda}\left(1+\frac{zF_p''(z)}{F_p'(z)}\right) = e^{i\lambda}p\left(1-\sum_{i=1}^{n}\alpha_i\right)+e^{i\lambda}\sum_{i=1}^{n}\alpha_i\frac{zf_i'(z)}{f_i(z)}.$$

$$(2.3)$$

Subtracting and adding $\rho\cos\lambda\sum_{i=1}^{n}\alpha_i$ on the right hand side of (2.3), we have

$$e^{i\lambda}\left(1+\frac{zF_p''(z)}{F_p'(z)}\right) = pe^{i\lambda}+\left(\rho\cos\lambda-pe^{i\lambda}\right)\sum_{i=1}^{n}\alpha_i$$

$$+\sum_{i=1}^{n}\alpha_i\left[e^{i\lambda}\frac{zf_i'(z)}{f_i(z)}-\rho\cos\lambda\right],$$

$$(2.4)$$

Taking real part of (2.4) and then simple computation gives

$$\int_0^{2\pi}\left|\mathrm{Re}\left[e^{i\lambda}\left(1+\frac{zF_p''(z)}{F_p'(z)}\right)-\lambda_1\cos\lambda\right]\right|d\theta$$

$$\leq\sum_{i=1}^{n}\alpha_i\int_0^{2\pi}\left|\mathrm{Re}\left[e^{i\lambda}\frac{zf_i'(z)}{f_i(z)}-\rho\cos\lambda\right]\right|d\theta,$$

$$(2.5)$$

where λ_1 is given by (2.1). Since $f_i(z)\in R_k^\lambda(p,\rho)$ for $1\leq i\leq n$, we have

$$\int_0^{2\pi}\left|\mathrm{Re}\left[e^{i\lambda}\frac{zf_i'(z)}{f_i(z)}-\rho\cos\lambda\right]\right|d\theta\leq(p-\rho)\cos\lambda k\pi.$$

$$(2.6)$$

Using (2.6) and (2.1) in (2.5), we obtain

$$\int_0^{2\pi}\left|\mathrm{Re}\left[e^{i\lambda}\left(1+\frac{zF_p''(z)}{F_p'(z)}\right)-\lambda_1\cos\lambda\right]\right|d\theta\leq(p-\lambda_1)\cos\lambda k\pi.$$

Hence $f_n(z) \in V_k^\lambda(p,\lambda_1)$ with λ_1 is given by (2.1).

By setting p=1 and $\lambda=0$ in Theorem 2.1, we obtain the following result proved in [9].

Corollory 2.2:

Let $f_i(z) \in R_k(p)$ for $1 \le i \le n$ with $0 \le p < 1$. Also let $\alpha_i > 0$, $1 \le i \le n$. If

$$0 \le (\rho-1)\sum_{i=1}^{n}\alpha_i + 1 < 1,$$

then $_n(z) \in V_k(\lambda_1)$ and λ_1 is given by (2.1).

Now if we take k=2 and $\lambda=0$ in Theorem 2.1, we obtain the following result.

Corollory 2.3:

Let $f_i(z) \in S_p^*(p)$ for $1 \le i \le n$ with $0 \le p < 1$. Also let $\alpha_i > 0, 1 \le i \le n$. If

$$0 \le (\rho-p)\sum_{i=1}^{n}\alpha_i + p < 1,$$

then $F_p(z) \in C_p(\lambda_1)$ and λ_1 is given by (2.1).

Letting $p=n=1; \lambda=0; \alpha_1 = \alpha$ and $f_1(z)=f(z)$ in Theorem 2.1, we have.

Corollory 2.4:

Let $f(z) \in R_k(\rho)$ with $0 \le \rho < 1$. Also let $\alpha > 0$. If

$$0 \le (\rho - 1)\alpha + 1 < 1,$$

then

$$\int_0^z \left(\frac{f(t)}{t} \right)^\alpha dt \in V_k(\lambda_1)$$

with $\lambda_1 = (\rho - 1)\alpha + 1$.

Theorem 2.5:

Let $f_i(z) \in V_k^\lambda(p, \rho)$ for $1 \le i \le n$

With $0 \le \rho < 1$. Also let λ is real is real with $|\lambda| < \frac{\pi}{2}$,

$\alpha_i > 0, 1 \le i \le n$. If

$$0 \le (\rho - p) \sum_{i=1}^n \alpha_i + p < 1,$$

then $G_p(z) \in V_k^\lambda(p, \lambda_1)$ and λ_1 is given by (2.1).

Proof: From (1.2), we have

$$1 + \frac{zG_p''(z)}{G_p'(z)} = p + \sum_{i=1}^n \alpha_i \left(\frac{zf_i''(z)}{f_i'(z)} + 1 \right) - p \sum_{i=1}^n \alpha_i,$$

or, equivalently

$$e^{i\lambda}\left(1+\frac{zG_p''(z)}{G_p'(z)}\right)$$

$$= pe^{i\lambda}\left(1-\sum_{i=1}^{n}\alpha_i\right)+\sum_{i=1}^{n}\alpha_i e^{i\lambda}\left(1+\frac{zf_i''(z)}{f_i'(z)}\right).$$

This relation is equivalent to

$$e^{i\lambda}\left(1+\frac{zG_p''(z)}{G_p'(z)}\right)=pe^{i\lambda}+\left(\rho\cos\lambda-pe^{i\lambda}\right)\sum_{i=1}^{n}\alpha_i$$

$$+\sum_{i=1}^{n}\alpha_i\left[e^{i\lambda}\left(1+\frac{zf_i''(z)}{f_i'(z)}\right)-\rho\cos\lambda\right].$$

$$(2.7)$$

Taking real part of (2.7) and then simple computation gives us

$$\int_{0}^{2\pi}\left|\text{Re}\left[e^{i\lambda}\left(1+\frac{zG_p''(z)}{G_p'(z)}\right)-\lambda_1\cos\lambda\right]\right|d\theta$$

$$\leq\sum_{i=1}^{n}\alpha_i\int_{0}^{2\pi}\left|\text{Re}\left[e^{i\lambda}\left(1+\frac{zf_i''(z)}{f_i'(z)}\right)-\rho\cos\lambda\right]\right|d\theta,$$

$$(2.8)$$

where λ_1 is given by (2.1). Since $f_i(z)\in V_k^\lambda(p,\rho)$ for $1\leq i\leq n$, we have

$$\int_{0}^{2\pi}\left|\text{Re}\left[e^{i\lambda}\left(1+\frac{zf_i''(z)}{f_i'(z)}\right)-\rho\cos\lambda\right]\right|d\theta\leq(p-\rho)\cos\lambda k\pi.$$

$$(2.9)$$

Using (2.9) in (2.8), we obtain

$$\int_0^{2\pi} \left| Re\left[e^{i\lambda} \left(1 + \frac{zG_p''(z)}{G_p'(z)} \right) - \lambda_1 \cos\lambda \right] \right| d\theta \le (p - \lambda_1)\cos\lambda k\pi.$$

Hence $G_p(z) \in V_k^\lambda(p,\lambda_1)$ with λ_1 is given by (2.1).

By setting k=2 and $\lambda=0$ in Theorem 2.5, we obtain the following result.

Corollory 2.6:

Let $f_i(z) \in C_p(\rho)$ for $1 \le i \le n$ with $0 \le \rho < 1$. Also let $\alpha_i > 0, 1 \le i \le n$. If

$$0 \le (\rho - p)\sum_{i=1}^{n}\alpha_i + p < 1,$$

then $G_p(z) \in C_p(\lambda_1)$ with λ_1 is given by (2.1).

Letting p=n=1, $\lambda=0$, $\alpha_1 = \delta$ and $f_1(z) = f(z)$ in Theorem 2.5, we have.

Corollory 2.7:

Let $f(z) \in V_k(\rho)$ with $0 \le \rho < 1$. Also let $\delta > 0$. If $0 \le (\rho-1)\delta+1 < 1$, then

$$\int_0^z (f'(t))^\delta \, dt \in V_k(\lambda_1)$$

with

$$\lambda_1 = (\rho - 1)\delta + 1.$$

REFERENCES

1. L. Spacek, "Prispěvek k teorii funkei prostych," Časopis pro pěstováni matematiky a fysiky, Vol. 62, No. 2, 1933, pp. 12-19.

2. M. S. Robertson, "Univalent functions f(z) for wich zf'(z) is spiral-like," Michigan Mathematical Journal, Vol. 16, No. 2, 1969, pp. 97-101.

3. K. S. Padmanabhan and R. Parvatham, "Properties of a class of functions with bounded boundary rotation," Annales Polonici Mathematici, Vol. 31, No. 1, 1975, pp. 311-323.

4. B. Pinchuk, "Functions with bounded boundary rotation," Israel Journal of Mathematics, Vol. 10, No. 1, 1971, pp. 7-16. doi:10.1007/BF02771515

5. O. Tammi, "On the maximization of the coefficients of Schlicht and related functions," Annales Academiae Scientiarum Fennicae. Series A I. Mathematica, Vol. 114, No. 1, 1952, p. 51

6. V. Paatero, "Uber Gebiete von beschrankter Randdrehung," Annales Academiae Scientiarum Fennnicae, Vol. 37-39, No. 9, 1933.

7. M. Arif, M. Ayaz and S. I. Ali Shah, "Radii problems for certain classes of analytic functions with fixed second coefficients," World Applied Sciences Journal, Vol. 13, No. 10, 2011, pp. 2240-2243.

8. K. I. Noor, "On some subclasses of fuctions with bounded boundary and bounded radius rotation," Pan American Mathematical Journal, Vol. 6, No. 1, 1996, pp. 75-81.

9. K. I. Noor, M. Arif and W. Haq, "Some properties of certain integral opertors," Acta Universitatis Apulensis, Vol. 21, 2010, pp. 89-95.

10. K. I. Noor, M. Arif and A. Muhammad, "Mapping properties of some classes of analytic functions under an integral operator," Journal of Mathematical Inequalities, Vol. 4, No. 4, 2010, pp. 593-600.

11. K. I. Noor, W. Haq, M. Arif and S. Mustafa, "On bounded boundary and bounded radius rotations," Journal of Inequalities and Applications, 2009, Article ID: 813687.

12. K. I. Noor, S. N. Malik, M. Arif and M. Raza, "On bounded boundary and bounded radius rotation related with Janowski function," World Applied Sciences Journal, Vol. 12, No. 6, 2011, pp. 895-902.

13. B. A. Frasin, "New general integral operators of p-valent functions," Journal of Inequatilies Pure and Applied Mathematics, Vol. 10, No. 4, 2009

14. D. Breaz and N. Breaz, "Two Integral Operators," Studia Universitatis Babes-Bolyai, Mathematica, Clunj-Napoca, Vol. 47, No. 3, 2002, pp. 13-21.

15. D. Breaz, S. Owa and N. Breaz, "A New Integral Univalent Operator," Acta Universitatis Apulensis, Vol. 16, 2008, pp. 11-16.

16. N. Breaz, V. Pescar and D. Breaz, "Univalence criteria for a new integral operator," Mathematical and Computer Modelling, Vol. 52, No. 1-2, 2010, pp. 241-246.doi:10.1016/j.mcm.2010.02.013

17. B. A. Frasin, "Convexity of integral operators of p-valent functions," Mathematical and Computer Modelling, Vol. 51, No. 5-6, 2010, pp. 601-605.

18. B. A. Frasin, "Some sufficient conditions for certain integral operators," Journal of Mathematics and Inequalities, Vol. 2. No. 4, 2008, pp. 527-535.

19. G. Saltik, E. Deniz and E. Kadioglu, "Two new general p-valent integral operators," Mathematical and Computer Modelling, Vol. 52, No. 9-10, 2010, pp. 1605-1609.doi:10.1016/j.mcm.2010.06.025

20. R. M. Ali and V. Ravichandran, "Integral operators on Ma-Minda type starlike and convex functions," Mathematical and Computer Modelling, Vo. 53, No. 5-6, 2011, pp. 581-586.doi:10.1016/j.mcm.2010.09.007

21. S. S. Miller, P. T. Mocanu and M. O. Reade, "Starlike integral operators," Pacific Journal of Mathematics, Vol. 79, No. 1, 1978, pp.157-168.

22. Y. J. Kim and E. P. Merkes, "On an integral of powers of a spirallike function," Kyungpook Mathematical Journal, Vol. 12, No. 2, 1972, pp. 249-252.

23. N. N. Pascu and V. Pescar, "On the integral operators of Kim-Merkes and Pfaltz-graff," Mathematica, Universitatis Babes-Bolyai Cluj-Napoca, Vol. 32, No. 2, 1990, pp. 185- 192.

CITATION

M. Arif, W. Ul-Haq and M. Ismail, "Mapping Properties of Generalized Robertson Functions under Certain Integral Operators," Applied Mathematics, Vol. 3 No. 1, 2012, pp. 52-55. doi: 10.4236/am.2012.31009.

Some Applications of Fractional Calculus in Engineering

J. A. Tenreiro Machado, Manuel F. Silva, Ramiro S. Barbosa, Isabel S. Jesus, Cecília M. Reis, Maria G. Marcos, and Alexandra F. Galhano
Institute of Engineering of Porto, Porto, Portugal

ABSTRACT

Fractional Calculus (FC) goes back to the beginning of the theory of differential calculus. Nevertheless, the application of FC just emerged in the last two decades, due to the progress in the area of chaos that revealed subtle relationships with the FC concepts. In the field of dynamical systems theory some work has been carried out but the proposed models and algorithms are still in a preliminary stage of establishment. Having these ideas in mind, the paper discusses FC in the study of system dynamics and control. In this perspective, this paper investigates the use of FC in the fields of controller tuning, legged robots, redundant robots, heat diffusion, and digital circuit synthesis.

INTRODUCTION

The generalization of the concept of derivative $D^{\alpha}[f(x)]$ to noninteger values of α goes back to the beginning of the theory of differential calculus. In fact, Leibniz, in his correspondence with Bernoulli, L'Hôpital and Wallis (1695), had several notes about the calculation of $D^{1/2}[f(x)]$. Nevertheless, the development of the theory of Fractional Calculus

(FC) is due to the contributions of many mathematicians such as Euler, Liouville, Riemann, and Letnikov [1–3].

The FC deals with derivatives and integrals to an arbitrary order (real or, even, complex order). The mathematical definition of a derivative/ integral of fractional order has been the subject of several different approaches [1–3]. For example, the Laplace definition of a fractional derivative of a signal $x(t)$ is

$$D^\alpha x(t) = L^{-1}\left\{ s^\alpha X(s) - \sum_{k=0}^{n-1} s^k D^{\alpha-k-1} x(t)|_{t=0} \right\},$$

(1.1)

where , $n-1 < \alpha \le n, \alpha > 0$. The Grünwald-Letnikov definition is given by $(\alpha \in \Re)$:

$$D^\alpha x(t) = \lim_{h \to 0} \left[\frac{1}{h^\alpha} \sum_{k=0}^{\infty} (-1)^k \binom{\alpha}{k} x(t - kh) \right],$$

$$\binom{\alpha}{k} = \frac{\Gamma(\alpha + 1)}{\Gamma(k + 1)\Gamma(\alpha - k + 1)},$$

(1.2)

where Γ is the Gamma function and h is the time increment. However, (1.2) shows that fractional-order operators are "global" operators having a memory of all past events, making them adequate for modeling memory effects in most materials and systems.

The Riemann-Liouville definition of the fractional-order derivative is $(\alpha > 0)$:

$$_aD_t^\alpha f(t) = \frac{1}{\Gamma(n - \alpha)} \frac{d^n}{dt^n} \int_a^t \frac{f(\tau)}{(t - \tau)^{\alpha-n+1}} d\tau, \quad n - 1 < \alpha < n,$$

(1.3)

where $\Gamma(x)$ is the Gamma function of x.

Based on the proposed definitions it is possible to calculate the fractional-order integrals/derivatives of several functions (Table 1). Nevertheless, the problem of devising and implementing fractional-order algorithms is not trivial and will be the matter of the following sections.

Table 1: Fractional-order integrals of several functions

$\varphi(x)$, $x \in \mathfrak{R}$	$(I+\alpha\varphi)(x)$, $x \in \mathfrak{R}$, $\alpha \in C$
$(x-a)\beta-1$	$\Gamma(\beta)\Gamma(\alpha+\beta)(x-a)\alpha+\beta-1$, $Re(\beta)>0$
$e\lambda x$	$\lambda-\alpha e\lambda x$, $Re(\lambda)>0$
$\{\sin(\lambda x)\cos(\lambda x)$	$\lambda-\alpha\{\sin(\lambda x-\alpha\pi/2),\cos(\lambda x-\alpha\pi/2)$, $\lambda>0$, $Re(\alpha)>1$
$e\lambda x\{\sin(\gamma x)\cos(\gamma x)$	$e\lambda x(\lambda 2+\gamma 2)\alpha/2\{\sin(\gamma x-\alpha\varphi),\varphi=\arctan(\gamma/\lambda)\cos(\gamma x-\alpha\varphi),\gamma>0$, $Re(\lambda)>1$

In recent years FC has been a fruitful field of research in science and engineering [1–6]. In fact, many scientific areas are currently paying attention to the FC concepts and we can refer its adoption in viscoelasticity and damping, diffusion and wave propagation, electromagnetism, chaos and fractals, heat transfer, biology, electronics, signal processing, robotics, system identification, traffic systems, genetic algorithms, percolation, modeling and identification, telecommunications, chemistry, irreversibility, physics, control systems as well as economy, and finance [7–18].

Bearing these ideas in mind, Sections 2–6 present several applications of FC in science and engineering. In Section 2, it is presented the application of FC concepts to the tuning of PID controllers and, in Section 3, the application of a fractional-order PD controller in the control of the leg joints of a hexapod robot. Then in Section 4, it is presented the fractional dynamics in the trajectory control of redundant manipulators. Next, in Section 5, it is introduced the fractional characteristics of heat diffusion along a media and, in Section 6 it is shown the application of FC to circuit synthesis using evolutionary algorithms. Finally, the main conclusions are presented in Section 7.

TUNING OF PID CONTROLLERS USING FRACTIONAL CALCULUS CONCEPTS

The PID controllers are the most commonly used control algorithms in industry. Among the various existent schemes for tuning PID controllers, the Ziegler-Nichols (Z-N) method is the most popular and is still extensively used for the determination of the PID parameters. It is well known that the compensated systems, with controllers tuned by this method, have generally a step response with a high percent overshoot. Moreover, the Z-N heuristics are only suitable for plants with monotonic step response.

In this section, we study a methodology for tuning PID controllers such that the response of the compensated system has an almost constant overshoot defined by a prescribed value. The proposed method is based on the minimization of the integral of square error (ISE) between the step responses of a unit feedback control system, whose open-loop transfer function L(s) is given by a fractional-order integrator and that of the PID compensated system [7].

Figure 1 illustrates the fractional-order control system that will be used as reference model for the tuning of PID controllers. The open-loop transfer function L(s) is defined as $(\alpha \in R^+)$:

$$L(s) = \left(\frac{\omega_c}{s}\right)^{\alpha}, \tag{2.1}$$

where ω_c is the gain crossover frequency, that is, $|L(j\omega_c)| = 1$. The parameter α is the slope of the magnitude curve, on a log-log scale, and may assume integer as well as noninteger values. In this study we consider $1 < \alpha < 2$, such that the output response may have a fractional oscillation (similar to an underdamped second-order system). This transfer function is also known as the Bode's ideal loop transfer function since Bode studies on the design of feedback amplifiers in the 1940s [19].

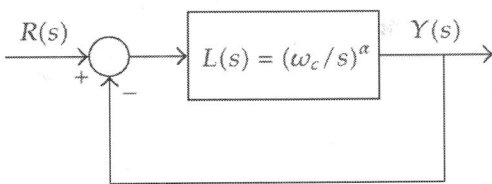

Figure 1: Fractional-order control system with open-loop transfer function L(s).

The Bode diagrams of amplitude and phase of L(s) are illustrated in Figure 2. The amplitude curve is a straight line of constant slope $-20\alpha \text{dB/dec}$, and the phase curve is a horizontal line positioned at $-\alpha\pi/2$ rad. The Nyquist curve is simply the straight line through the origin, arg $L(j\omega) = -\alpha\pi/2$ rad.

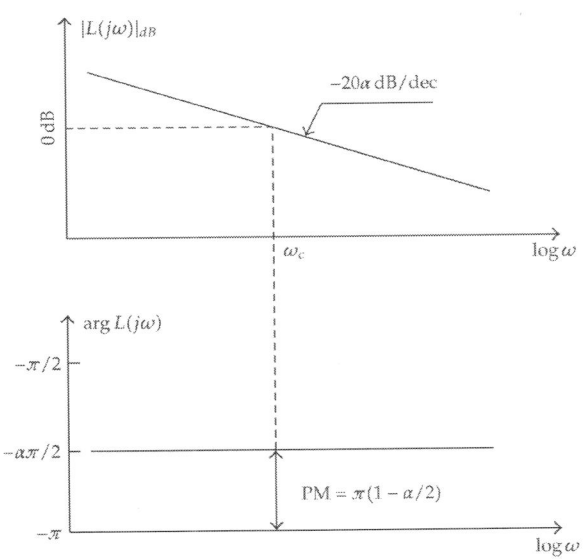

Figure 2: Bode diagrams of amplitude and phase of $L(j\omega)$ for $1 < \alpha < 2$.

This choice of L(s) gives a closed-loop system with the desirable property of being insensitive to gain changes. If the gain changes, the crossover frequency ω_c will change, but the phase margin of the system remains $PM = \pi(1 - \alpha/2)$ rad, independent of the value of the gain. This can be seen from the curves of amplitude and phase of Figure 2.

The closed-loop transfer function of fractional-order control system of Figure 1 is given by

$$G(s) = \frac{L(s)}{1 + L(s)} = \frac{1}{(s/\omega_c)^\alpha + 1}, \quad 1 < \alpha < 2.$$

(2.2)

The unit step response of G(s) is given by the expression:

$$y_d(t) = L^{-1}\left\{\frac{1}{s}G(s)\right\} = L^{-1}\left\{\frac{\omega_c^\alpha}{s(s^\alpha + \omega_c^\alpha)}\right\} = 1 - \sum_{n=0}^{\infty}\frac{[-(\omega_c t)^\alpha]^n}{\Gamma(1 + \alpha n)} = 1 - E_\alpha[-(\omega_c t)^\alpha].$$

(2.3)

For the tuning of PID controllers, we address the fractional-order transfer function (2.2) as the reference system [8]. With the order α and the crossover frequency ω_c we can establish the overshoot and the speed of the output response, respectively. For that purpose we consider the closed-loop system shown in Figure 3, where $G_c(s)$ and $G_p(s)$ are the PID controller and the plant transfer functions, respectively.

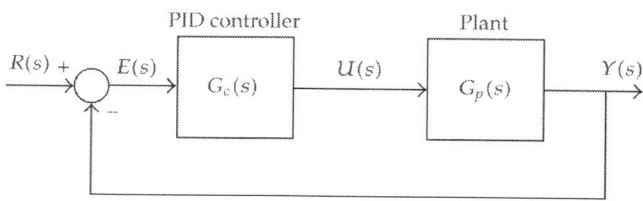

Figure 3: Closed-loop control system with PID controller $G_c(s)$.

The transfer function of the PID controller is

$$G_c(s) = \frac{U(s)}{E(s)} = K\left(1 + \frac{1}{T_i s} + T_d s\right),$$

(2.4)

where $E(s)$ is the error signal and $U(s)$ is the controller's output. The parameters K, T_i and T_d are the proportional gain, the integral time constant, and the derivative time constant of the controller, respectively.

The design of the PID controller will consist on the determination of the optimum PID set gains (K, T_i, T_d) that minimize J, the integral of the square error (ISE), defined as

$$J = \int_0^\infty [y(t) - y_d(t)]^2 dt,$$

(2.5)

where $y(t)$ is the step response of the closed-loop system with the PID controller (Figure 3) and $y_d(t)$ is the desired step response of the fractional-order transfer function (2.2) given by (2.3).

To illustrate the effectiveness of proposed methodology we consider the third-order plant transfer function:

$$G_p(s) = \frac{K_p}{(s + 1)^3}$$

(2.6)

with nominal gain $K_p = 1$.

Figure 4 shows the step responses and the Bode diagrams of phase of the closed-loop system with the PID for the transfer function $G_p(s)$ for gain variations around the nominal gain $(K_p = 1)$ corresponding to $K_p = \{0.6, 0.8, 1.0, 1.2, 1.4\}$ that is, for a variation up to $\pm 40\%$ of its nominal value. The system was tuned for $\alpha = 3/2 (PM = 45°)$, $\omega_c = 0.8$ rad/s. We verify that we get the same desired iso-damping property corresponding to the prescribed (α, ω_c) values.

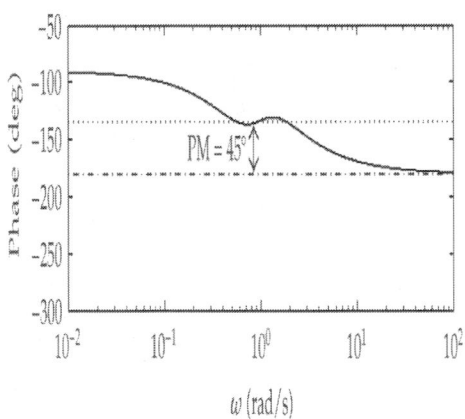

(a)

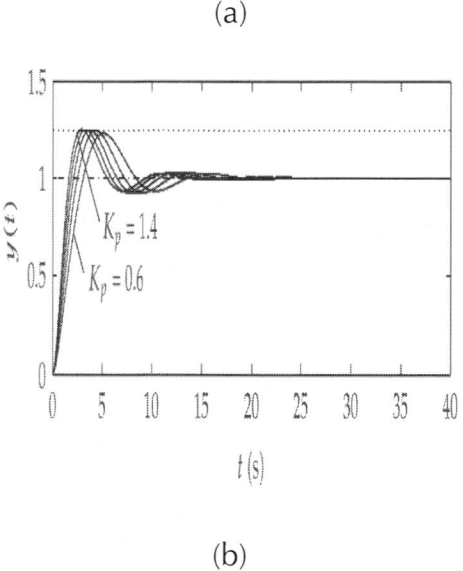

(b)

Figure 4: Bode phase diagrams and step responses for the closed-loop system with a PID controller for $G_p(s)$. The PID parameters are K=1.9158, T_i=1.1407, and T_d=0.9040.

In fact, we observe that the step responses have an almost constant overshoot independent of the variation of the plant gain around the gain crossover frequency ω_c. Therefore, the proposed methodology is capable of producing closed-loop systems robust to gain variations and step responses exhibiting an iso-damping property. The proposed method was tested on several systems revealing good results. It was also compared with other tuning methods showing comparable or superior results [8].

FRACTIONAL PD^{α} CONTROL OF A HEXAPOD ROBOT

Walking machines allow locomotion in terrain inaccessible to other type of vehicles, since they do not need a continuous support surface, but at the cost of higher requirements for leg coordination and control. For these robots, joint level control is usually implemented through a PID-like scheme with position feedback. Recently, the application of the theory of FC to robotics revealed promising aspects for future developments [9]. With these facts in mind, this section compares different Fractional PD^{α} robot controller tuning, applied to the joint control of a walking system (Figure 5) with n=6 legs, equally distributed along both sides of the robot body, having each three rotational joints $\left(\text{i.e.}, j=\{1,2,3\}\equiv\{\text{hip,knee,ankle}\}\right)$ [10].

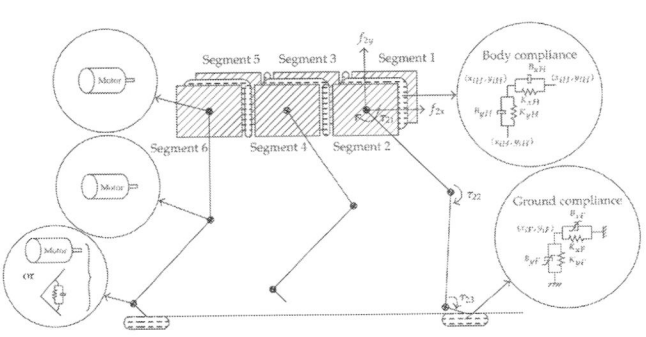

Figure 5: Model of the robot body and foot-ground interaction.

During this study leg joint j=3 can be either mechanical actuated or motor actuated (Figure 5). For the mechanical actuated case, we suppose that there is a rotational pre-tensioned spring-dashpot system connecting leg links L_{i2} and L_{i3}. This mechanical impedance maintains the angle between the two links while imposing a joint torque [10].

Figure 5 presents the dynamic model for the hexapod body and foot-ground interaction. It is considered robot body compliance because walking animals have a spine that allows supporting the locomotion with improved stability. The robot body is divided in n identical segments (each with mass $M_b n^{-1}$) and a linear spring-damper system (with parameters defined so that the body behaviour is similar to the one expected to occur on an animal) is adopted to implement the intrabody compliance [10]. The contact of the ith robot feet with the ground is modelled through a nonlinear system [11], being the values for the parameters based on the studies of soil mechanics [11].

The general control architecture of the hexapod robot is presented in Figure 6 [12]. In this study we evaluate the effect of different PD^{α}, $\alpha \in \Re$, controller implementations for $G_{c1}(s)$, while G_{c2} is a proportional controller with gain $Kp_j = 0.9 (j=1,2,3)$. For the PD^{α} algorithm, implemented through a discrete-time 4th-order Padé approximation $(a_{ij}, b_{ij} \in \Re, j=1,2,3)$, we have

$$G_{c1j}(z) \approx Kp_j + Ka_j \frac{\sum_{i=0}^{i=u} a_{ij}\, z^{-i}}{\sum_{i=0}^{i=u} b_{ij}\, z^{-i}}, \tag{3.1}$$

where Kp_j and Ka_j are the proportional and derivative gains, respectively, and α_j is the fractional order, for joint j. Therefore, the classical PD^1 algorithm occurs when the fractional order $\alpha_j = 1.0$.

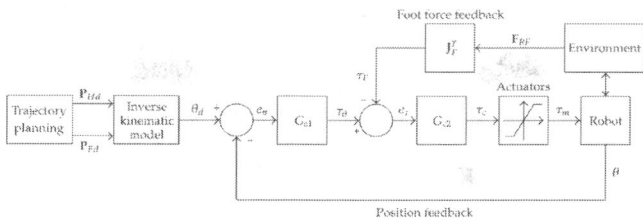

Figure 6: Hexapod robot control architecture.

It is analysed the system performance of the different PDᵅ tuning, during a periodic wave gait at a constant forward velocity V_F, for two cases: two leg joints are motor actuated and the ankle joint is mechanical actuated and the three leg joints are fully motor actuated [10].

The analysis is based on the formulation of two indices measuring the mean absolute density of energy per traveled distance (E_{av}) and the hip trajectory errors (ε_{xyH}) during walking, according to

$$E_{av} = \frac{1}{d}\sum_{i=1}^{n}\sum_{j=1}^{m}\int_0^T |\tau_{ij}(t)\dot{\theta}_{ij}(t)|\,dt \quad [Jm^{-1}],$$

$$\varepsilon_{xyH} = \sum_{i=1}^{n}\sqrt{\frac{1}{N_s}\sum_{k=1}^{N_s}\left(\Delta_{ixH}^2 + \Delta_{iyH}^2\right)} \quad [m],$$

$$\Delta_{ixH} = x_{iHd}(k) - x_{iH}(k), \qquad \Delta_{iyH} = y_{iHd}(k) - y_{iH}(k).$$

(3.2)

To tune the different controller implementations we adopt a systematic method, testing and evaluating several possible combinations of parameters, for all controller implementations. Therefore, we adopt the

$G_{c1}(s)$ parameters that establish a compromise in what concerns the simultaneous minimisation of E_{av} and (ε_{xyH}). Moreover, it is assumed high-performance joint actuators, with a maximum actuator torque of $\tau_{ijMax} = 400\,Nm$, and the desired angle between the foot and the ground (assumed horizontal) is established as $\theta_{i3hd}=-15°$. We tune the PD^α joint controllers for different values of the fractional order α_j while making $\alpha_1=\alpha_2=\alpha_3$.

We start by considering that leg joints 1 and 2 are motor actuated and joint 3 is mechanical actuated. For this case we tune the PD^α joint controllers for different values of the fractional order α_j, with step $\Delta\alpha_j=0.1$, namely, $\alpha_j=\{-09,-0.8,....,+0.9\}$. Afterwards, we consider that joint 3 is also motor actuated, and we repeat the controller tuning procedure versus α_j.

For the first situation under study, we verify that the value of $\alpha_j=0.6$ (Figure 7), being the gains of the PD^α controller $K_{p1}=2500, K_{\alpha1}=800, K_{p2}=300, K_{\alpha2}=100$ and the parameters of the mechanical spring-dashpot system for the ankle actuation $K_3=1, B_3=2$ presents the best compromise situation between the simultaneous minimisation of $\mathcal{E}_{xyH}$ and E_{av}.

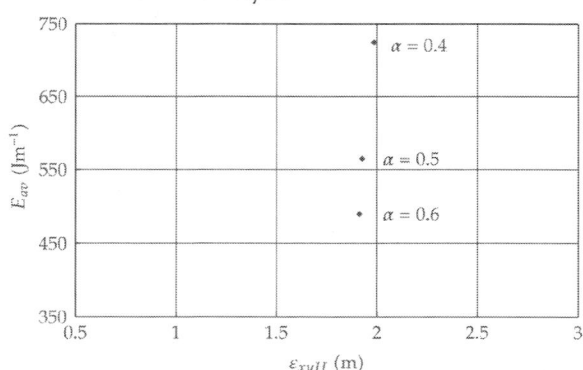

Figure 7: Locus of E_{av} versus $\mathcal{E}_{xyH}$ for the different values of α in the $G_{c1}(s)$ tuning, when establishing a compromise between the minimisation of E_{av} and $\mathcal{E}_{xyH}$, with $G_{c2}=0.9$, joints 1 and 2 motor actuated and joint 3 mechanical actuated.

Regarding the case when all joints are motor actuated, Figure 8 presents the best controller tuning for different values of α_j. The experiments reveal the superior performance of the PD^α controller for $\alpha_j \approx 0.5$, with $K_{p1}=15000, K_{\alpha1}=7200, K_{p2}=1000, K_{\alpha2}=800,$ and $K_{p3}=150, K_{\alpha3}=240$.

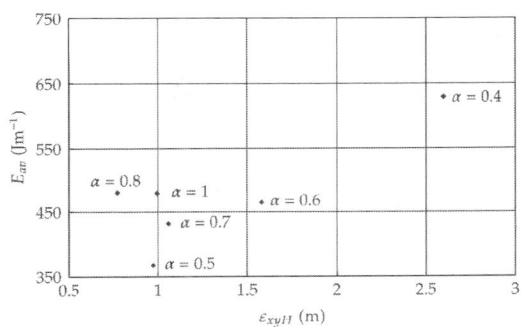

Figure 8: Locus of E_{av} versus ε_{xyH} for the different values of α in the $G_{c1}(s)$ tuning, when establishing a compromise between the minimisation of E_{av} and ε_{xyH}, with $G_{c2}=0.9$ and all joints motor actuated.

For $\alpha_j = \{0.1, 0.2, 0.3, 0.4\}$ the results are very poor and for $\alpha_j = \{-0.9,\ldots\ldots\ldots,-0.1\} \cup \{+0.9\}$, the hexapod locomotion is unstable. Furthermore, we conclude that the best case corresponds to all leg joints being motor actuated.

In conclusion, the experiments reveal the superior performance of the FO controller for $\alpha_i \approx 0.5$ and a robot with all motor actuated joints, as can be concluded analysing the curves for the joint actuation torques τ_{1jm} (Figure 9) and for the hip trajectory tracking errors Δ_{1xH} and Δ_{1yH} (Figure 10).

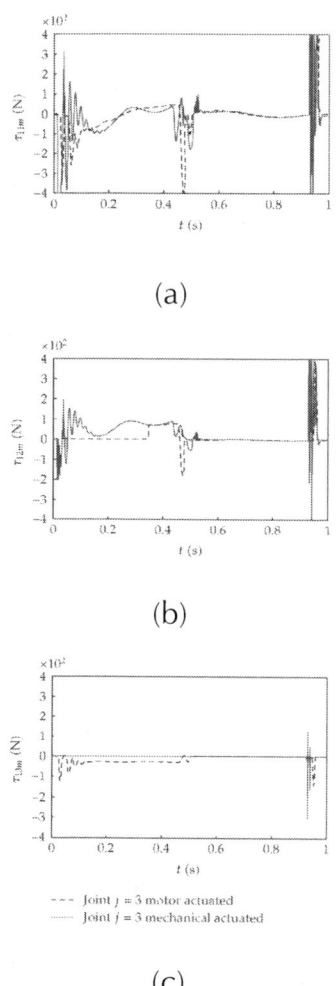

(a)

(b)

--- Joint $j = 3$ motor actuated
······ Joint $j = 3$ mechanical actuated

(c)

Figure 9: Plots of τ_{1jm} versus t, with joints 1 and 2 motor actuated and joint 3 mechanical actuated and all joints motor actuated, for $\alpha_j = 0.5$.

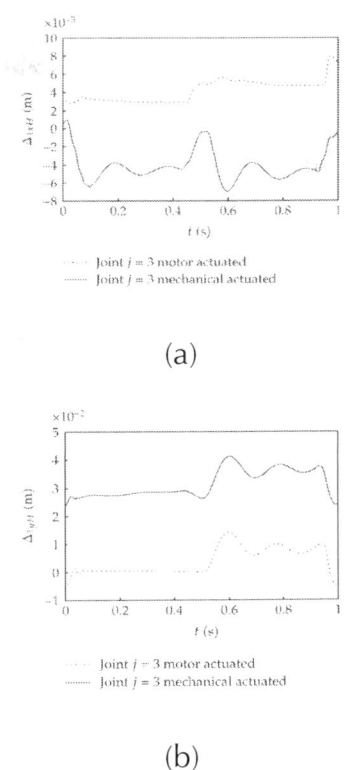

(a)

(b)

Figure 10: Plots of Δ_{1xH} and Δ_{1yH} versus t, with joints 1 and 2 motor actuated and joint 3 mechanical actuated and all joints motor actuated, for $\alpha_j = 0.5$.

Since the objective of the walking robots is to walk in natural terrains, in the sequel it is examined how the different controller tunings behave under different ground properties, considering that all joints are motor actuated. For this case, and considering the previously tuning controller parameters, the values of $\{K_{xF}, B_{xF}, K_{yF}, B_{yF}\}$ are varied simultaneously through a multiplying factor K_{mult} that is varied in the range [0.1,4.0]. This variation for the ground model parameters allows the simulation of the ground behaviour for growing stiffness, from peat to gravel [11].

The performance measure E_{av} versus the multiplying factor of the ground parameters K_{mult} is presented on Figure 11. Analysing the system performance from the viewpoint of the index E_{av}, it is possible to conclude that the best PD$^\alpha$ implementation occurs for the fractional order $\alpha_j = 0.5$. Moreover, it is clear that the performances of the different controller implementations are almost constant on all range of the ground parameters, with the exception of the fractional order $\alpha_j = 0.4$. For this case, E_{av} presents a significant variation with K_{mult}. Therefore, we conclude that the controller responses are quite similar, meaning that these algorithms are robust to variations of the ground characteristics [12].

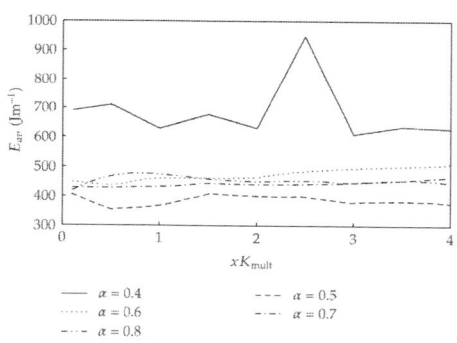

Figure 11: Performance index E_{av} versus K_{mult} for the $G_{c1}(s)$ PD$^\alpha$ controller tuning having all joints motor actuated.

FRACTIONAL DYNAMICS IN THE TRAJECTORY CONTROL OF REDUNDANT MANIPULATORS

A redundant manipulator is a robotic arm possessing more degrees of freedom (dof) than those required to establish an arbitrary position and orientation of the end effector. Redundant manipulators offer several potential advantages over non-redundant arms. In a workspace with obstacles, the extra degrees of freedom can be used to move around or between obstacles and thereby to manipulate in situations that otherwise would be inaccessible [20–23].

When a manipulator is redundant, it is anticipated that the inverse kinematics admits an infinite number of solutions. This implies that, for a given location of the manipulator's gripper, it is possible to induce a self-motion of the structure without changing the location of the end effecter. Therefore, the arm can be reconfigured to find better postures for an assigned set of task requirements.

Several kinematic techniques for redundant manipulators control the gripper through the rates at which the joints are driven, using the pseudoinverse of the Jacobian [22, 24]. Nevertheless, these algorithms lead to a kind of chaotic motion with unpredictable arm configurations.

Having these ideas in mind, Section 4.1 introduces the fundamental issues for the kinematics of redundant manipulators. Based on these concepts, Section 4.2 presents the trajectory control of a three dof robot. The results reveal a chaotic behavior that is further analyzed in Section 4.3.

Kinematics of Redundant Manipulators

A kinematically redundant manipulator has more dof than those required to define an arbitrary position and orientation of the gripper. In Figure 12 is depicted a planar manipulator with $k \in \aleph$ rotational (R) joints that is redundant for $k > 2$. When a manipulator is redundant it is anticipated that the inverse kinematics admits an infinite number of solutions. This implies that, for a given location of the manipulator's gripper, it is possible to induce a self-motion of the structure without changing the location of the gripper. Therefore, redundant manipulators can be reconfigured to find better postures for an assigned set of task requirements but, on the other hand, have a more complex structure requiring adequate control algorithms.

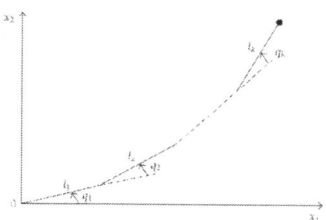

Figure 12: A planar redundant planar manipulator with
k rotational joints.

We consider a manipulator with n degrees of freedom whose joint variables are denoted by $q = [q_1, q_2, ..., q_n]^T$. We assume that a class of tasks, we are interested in can be described by m variables, $x = [x_1, x_2, ..., x_m]^T$ (m<n) and that the relation between q and x is given by

$$x = f(q),$$
(4.1)

where f is a function representing the direct kinematics.

Differentiating (4.1) with respect to time yields

$$\dot{x} = J(q)\dot{q},$$
(4.2)

where $\dot{x} \varepsilon \mathbb{R}^m$, $\dot{q} \varepsilon \mathbb{R}^n$ and $J(q) = \partial f(q)/\partial q \varepsilon \mathbb{R}^{m \times n}$. Hence, it is possible to calculate a path q(t) in terms of a prescribed trajectory x(t) in the operational space. We assume that the following condition is satisfied:

$$\max \text{rank}\{J(q)\} = m.$$
(4.3)

Failing to satisfy this condition usually means that the selection of manipulation variables is redundant and the number of these variables m can be reduced. When condition (4.3) is verified, we say that the degree of redundancy of the manipulator is n-m. If, for some q we have then the manipulator is in a singular state.

$$\text{rank}\{J(q)\} < m$$
(4.4)

This state is not desirable because, in this region of the trajectory, the manipulating ability is very limited.

Many approaches for solving redundancy [25, 26] are based on the inversion of (4.2). A solution in terms of the joint velocities is sought as

$$\dot{q} = J^{\#}(q)\dot{x}, \tag{4.5}$$

where $J^{\#}$ is one of the generalized inverses of the J [26–28]. It can be easily shown that a more general solution to (4.2) is given by

$$\dot{q} = J^{+}(q)\dot{x} + [I - J^{+}(q)J(q)]\dot{q}_0, \tag{4.6}$$

where I is the $n{\times}n$ identity matrix and $\dot{q}\varepsilon\mathbb{R}^n$ is a n×1 arbitrary joint velocity vector and J^+ is the pseudoinverse of the J. The solution (4.6) is composed of two terms. The first term is relative to minimum norm joint velocities. The second term, the homogeneous solution, attempts to satisfy the additional constraints specified by $\dot{q}_0$. Moreover, the matrix $I - J^+(q)J(q)$ allows the projection of $\dot{q}_0$ in the null space of J. A direct consequence is that it is possible to generate internal motions that reconfigure the manipulator structure without changing the gripper position and orientation [27–30]. Another aspect revealed by the solution of (4.6) is that repetitive trajectories in the operational space do not lead to periodic trajectories in the joint space. This is an obstacle for the solution of many tasks because the resultant robot configurations have similarities with those of a chaotic system.

Robot Trajectory Control

The direct kinematics and the Jacobian of a 3-link planar manipulator with rotational joints (3R robot) has a simple recursive nature according with the expressions:

$$\begin{bmatrix} x_1 \\ x_2 \end{bmatrix} = \begin{bmatrix} l_1 C_1 + l_2 C_{12} + l_3 C_{123} \\ l_1 S_1 + l_2 S_{12} + l_3 S_{123} \end{bmatrix},$$

$$J = \begin{bmatrix} -l_1 S_1 - \cdots - l_3 S_{123} & \cdots & -l_3 S_{123} \\ l_1 C_1 + \cdots + l_3 C_{123} & \cdots & l_3 C_{123} \end{bmatrix},$$

$$(4.7)$$

where l_i is the length of link $i, q_{i\ldots k} = q_i + \ldots + q_k, S_{i\ldots k} = Sin(q_{i\ldots k}),$ and $C_{i\ldots k} = Cos(q_{i\ldots k})..$

During all the experiments it is considered $\Delta t = 10^{-3}$ seconds, $L_{TOT} = l_1 = l_2 = l_3 = 3$ and $l_1 = l_2 = l_3$.

In the closed-loop pseudoinverse's method the joint positions can be computed through the time integration of the velocities according with the block diagram of the inverse kinematics algorithm depicted in Figure 13, where x_{ref} represents the vector of reference coordinates of the robot gripper in the operational space.

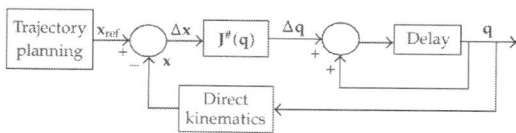

Figure 13: Block diagram of the closed-loop inverse kinematics algorithm with the pseudoinverse.

Based on (4.7) we analyze the kinematic performances of the 3R-robot when repeating a circular motion in the operational space with frequency $\omega_0 = 7.0 \, rads^{-1}$, centre at distance $r = [x_1^2 + x_2^2]^{1/2}$ and radius ρ.

Figure 14 shows the joint positions for the inverse kinematic algorithm (4.5) for $r = \{0.6, 2.0\}$ and $\rho = \{0.3, 0.5\}$. We observe that the following hold.

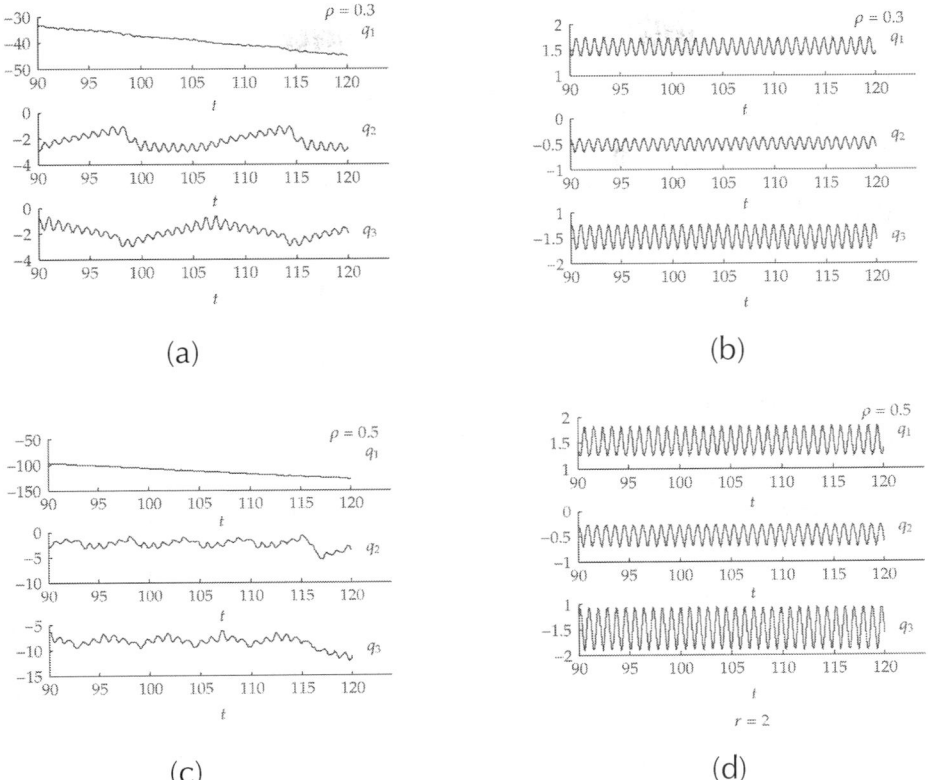

Figure 14: The 3R-robot joint positions versus time using the pseudo-inverse method for $r=\{0.6,2.0\}$ and $\rho=\{0.3,0.5\}$.

i. For $r=0.6$ occur unpredictable motions with severe variations that lead to high joint transients [13]. Moreover, we verify a low-frequency signal modulation that depends on the circle being executed.

ii. For $r=2.0$ the motion is periodic with frequency identical to $\omega_0 = 7.0\,\text{rads}^{-1}$.

Analysis of the Robot Trajectories

In the previous subsection we verified that the pseudoinverse-based algorithm leads to unpredictable arm configurations. In order to gain further insight into the pseudoinverse nature several distinct experiments are devised in the sequel during a time window of 300 cycles. Therefore, in a first set of experiments we calculate the Fourier transform of the 3R-robot joints velocities for a circular repetitive motion with frequency $\omega_0 = 7.0\,\mathrm{rads^{-1}}$, radius $\rho = \{0.1, 0.3, 0.5, 0.7\}$, and radial distances $r \varepsilon\,]0, L_{TOT} - \rho[$.

Figure 15 shows $|F\{\dot{q}_2(t)\}|$ versus the frequency ratio ω_0/ω and the distance r where $F\{\}$ represents the Fourier operator. Is verified an interesting phenomenon induced by the gripper repetitive motion because a large part of the energy is distributed along several subharmonics. These fractional-order harmonics (foh) depend on r and ρ making a complex pattern with similarities with those revealed by chaotic systems. Furthermore, we observe the existence of several distinct regions depending on r.

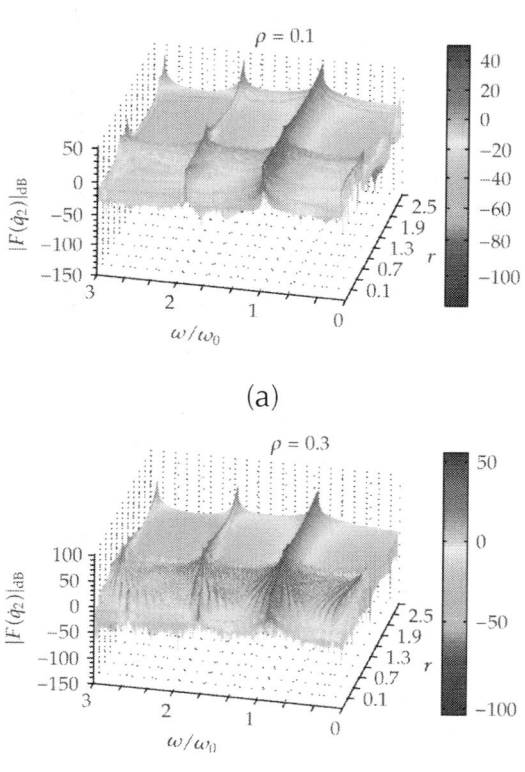

(a)

(b)

Some Applications of Fractional Calculus in Engineering

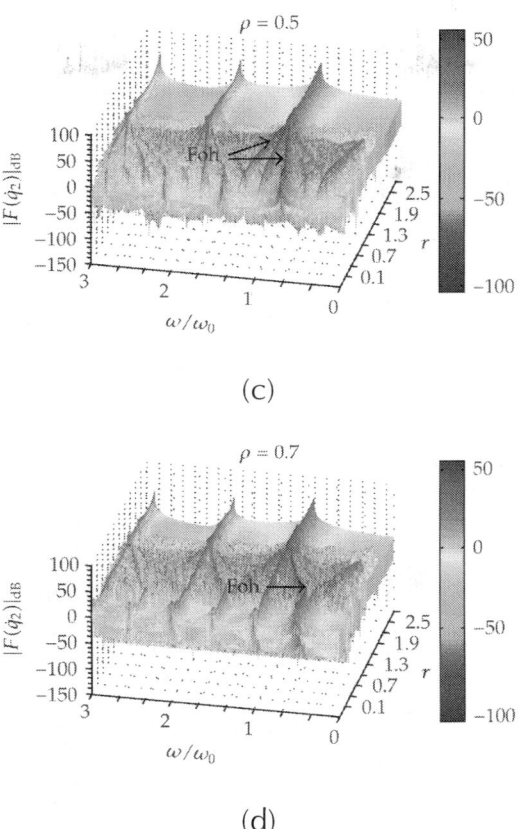

(c)

(d)

Figure 15: $\left|F\{\dot{q}_2(t)\}\right|$ of the 3R-robot during 300 cycles versus r and ω/ω_0, for $\rho=\{0.1,0.3,0.5,0.7\}$, $\omega_0=7.0\,\text{rads}^{-1}$.

For example, selecting in Figure 15 several distinct cases, namely for r={0.08,0.30,0.53,1.10,1.30,2.00}, we have the different signal Fourier spectra clearly visible in Figure 16.

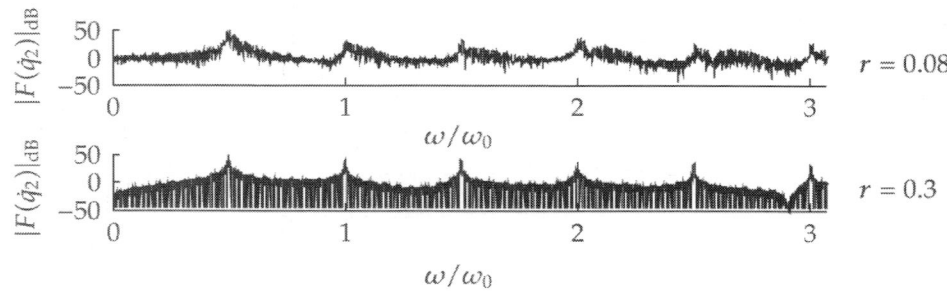

(a)

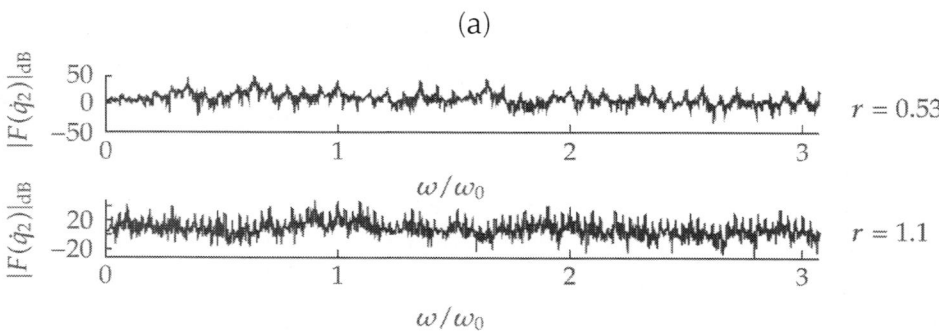

(b)

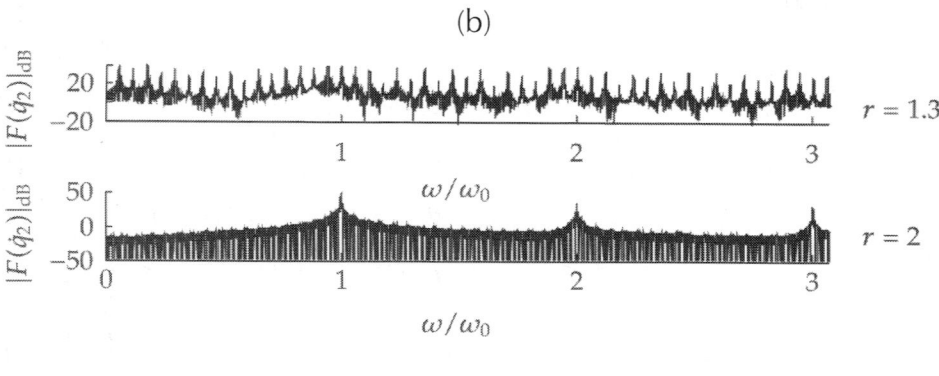

(c)

Figure 16: $\left|F\{\dot{q}_2(t)\}\right|$ of the 3R-robot during 300 cycles versus the frequency ratio ω_0/ω, for r={0.08,0.30,0.53,1.10,1.30,2.00}, $\rho=0.7$, $\omega_0=7.0\,\text{rads}^{-1}$.

In the authors' best knowledge the foh are aspects of fractional dynamics [14, 15, 31], but a final and assertive conclusion about a physical interpretation is a matter still to be explored.

For joints velocities 1 and 3 the results are similar to the verified ones for joint velocity 2.

HEAT DIFFUSION

The heat diffusion is governed by a linear one-dimensional partial differential equation (PDE) of the form:

$$\frac{\partial u}{\partial t} = k\frac{\partial^2 u}{\partial x^2},$$

$$(5.1)$$

where k is the diffusivity, t is the time, u is the temperature, and x is the space coordinate. However, (5.1) involves the solution of a PDE of parabolic type for which the standard theory guarantees the existence of a unique solution [16].

For the case of a planar perfectly isolated surface we usually apply a constant temperature U_0 at $x=0$ and analyzes the heat diffusion along the horizontal coordinate x. Under these conditions, the heat diffusion phenomenon is described by a noninteger-order model:

$$U(x,s) = \frac{U_0}{s}G(s) \qquad G(s) = e^{-x\sqrt{s/k}},$$

$$(5.2)$$

where x is the space coordinate, U_0 is the boundary condition, and G(s) is the system transfer function.

In our study, the simulation of the heat diffusion is performed by adopting the Crank-Nicholson implicit numerical integration based on the discrete approximation to differentiation as [16, 17]

$$-ru[j+1,i+1] + (2+r)u[j+1,i] - ru[j+1,i-1] = ru[j,i+1] + (2-r)u[j,i] + u[j,i-1],$$

(5.3)

where $r = \Delta t(\Delta x^2)^{-1}, \{\Delta x, \Delta t\}$ and {i,j} are the increments and the integration indices for space and time, respectively.

Control Strategies

The generalized PID controller $G_c(s)$ has a transfer function of the form

$$G_c(s) = K\left[1 + \frac{1}{T_i s^\alpha} + T_d s^\beta\right],$$

(5.4)

where α and β are the orders of the fractional integrator and differentiator, respectively. The constants K, T_i, and T_d are correspondingly the proportional gain, the integral time constant, and the derivative time constant.

Clearly, taking (α,β)={(1,1),(1,0),(0,1),(0,0)} we get the classical {PID,PI,PD,P} controllers, respectively.

The $PI^\alpha D^\beta$ controller is more flexible and gives the possibility of adjusting more carefully the closed-loop system characteristics.

In the following two subsections, we analyze the system of Figure 17 by adopting the classical integer-order PID and a fractional PID^β, respectively.

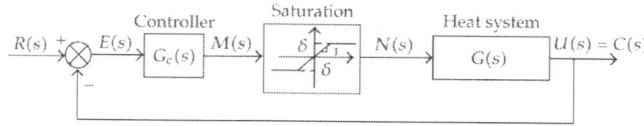

Figure 17: Closed-loop system with PID controller $G_c(s)$.

PID Tuning Using the Ziegler-Nichols Rule

In this subsection, we analyze the closed-loop system with a conventional PID controller given by the transfer function (5.4) with $\alpha = \beta = 1$. Usually, the PID parameters (K, T_i, T_d) are tuned by using the so-called Ziegler-Nichols open loop (ZNOL) method [17]. The ZNOL heuristics are based on the approximate first-order plus dead-time model:

$$\hat{G}(s) = \frac{K_p}{\tau s + 1} e^{-sT}.$$

$$(5.5)$$

For the heat system, the resulting parameters are $\{K_p, \tau, T\} = \{0.52, 162, 28\}$ leading to the PID constants $\{K, T_i, T_d\} = \{18.07, 34.0, 8.5\}$.

A step input is applied at x=0.0m and the closed-loop response c(t) is analyzed for x=3.0m, without actuator saturation (Figure 18). We verify that the system with a PID controller, tuned through the ZNOL heuristics, does not produce satisfactory results giving a significant overshoot ov and a large settling time t_s, namely $\{t_s, t_p, t_r, ov(\%)\} \equiv \{44.8, 27.5, 12.0, 68.56\}$ where t_p represents the peak time and t_r the rise time. We consider two indices that measure the response error, namely, the integral square error (ISE) and the integral time square error (ITSE) criteria defined as

$$ISE = \int_0^\infty [r(t) - c(t)]^2 dt,$$

$$ITSE = \int_0^\infty t[r(t) - c(t)]^2 dt.$$

$$(5.6)$$

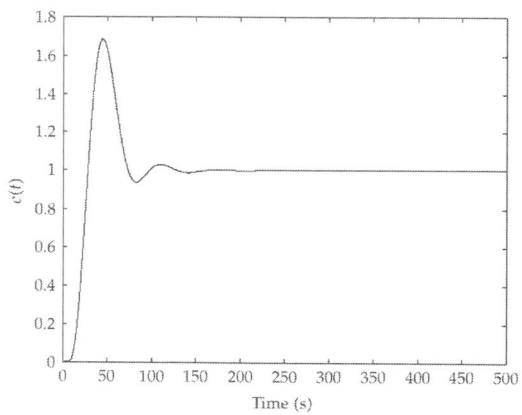

Figure 18: Step responses of the closed-loop system for the PID controller and x=3.0 m.

We can use other performance criteria such as the integral absolute error (IAE) or the integral time absolute error (ITAE); however, in the present case, the ISE and the ITSE criteria have produced the best results and are adopted in the study.

In this case, the ZNOL PID tuning leads to the values (ISE, ITSE)=(27.53,613.97). The poor results indicate again that the method of tuning may not be the most adequate for the control of the heat system.

In fact, the inherent fractional dynamics of the system lead us to consider other configurations. In this perspective, we propose the use of fractional controllers tuned by the minimization of the indices ISE and ITSE.

PID$^\beta$ Tuning Using Optimization Indices

In this subsection, we analyze the closed-loop system under the action of the PID$^\beta$ controller given by the transfer function (5.4) with $\alpha = 1$ and $0 \le \beta \le 1$. The fractional derivative term $T_d s^\beta$ in (5.4) is implemented through a fourth-order Padé discrete rational transfer function. It used a sampling period of T=0.1 second.

The PID$^\beta$ controller is tuned by the minimization of an integral performance index. For that purpose, we adopt the ISE and ITSE criteria.

A step reference input R(s)=1/s is applied at x=0.0 m and the output c(t) is analyzed for x=3.0m, without actuator saturation. The heat system is simulated for 3000 seconds. Figure 19 illustrates the variation of the fractional PID parameters (K,T$_i$,T$_d$) as function of the order's derivative β, for the ISE and the ITSE criteria. The dots represent the values corresponding to the classical PID (ZNOL-tuning) addressed in the previous section.

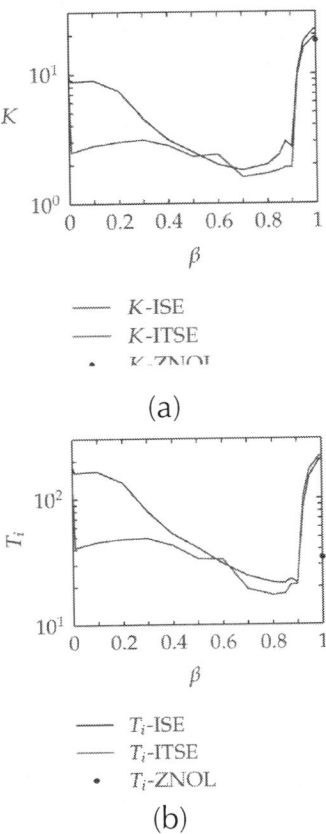

(a)

(b)

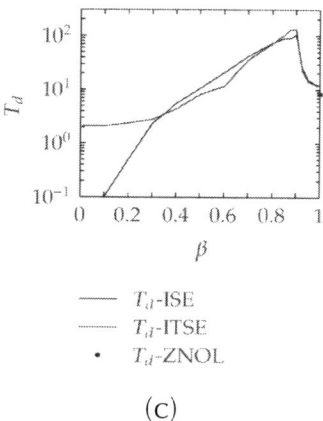

(c)

Figure 19: The PID$^\beta$ parameters (K, T_i, T_d) versus β for the ISE and ITSE optimization criteria. The dot represents the PID-ZNOL.

The curves reveal that for $\beta < 0.4$ the parameters (K, T_i, T_d) are slightly different, for the two ISE and ITSE criteria, while for $\beta \geq 0.4$ they lead to almost similar values. This fact indicates a large influence of a weak-order derivative on system's dynamics.

To further illustrate the performance of the fractional-order controllers a saturation nonlinearity is included in the closed-loop system of Figure 17 and inserted in series with the output of the controller $G_c(s)$. The saturation element is defined as

$$n(m) = \begin{cases} m, & |m| < \delta, \\ \delta \, \mathrm{sign}\,(m), & |m| \geq \delta. \end{cases} \tag{5.7}$$

The controller performance is evaluated for $\delta = \{20, \ldots, 100\}$ and $\delta = \infty$ which corresponds to a system without saturation. We use the same fractional-PID parameters obtained without considering the saturation nonlinearity.

Figures 20 and 21 show the step responses of the closed-loop system and the corresponding controller output, for the PID$^\beta$ tuned in the ISE and ITSE perspectives for $\delta=10$ and $\delta=\infty$, respectively. The controller parameters $\{K,T_i,T_d,\beta\}$ correspond to the minimization of those indices leading to the values ISE: $\{K,T_i,T_d,\beta\}\equiv\{3,23,90.6,0.875\}$ and ITSE: $\{K,T_i,T_d,\beta\}\equiv\{1.8,17.6,103.6,0.85\}$.

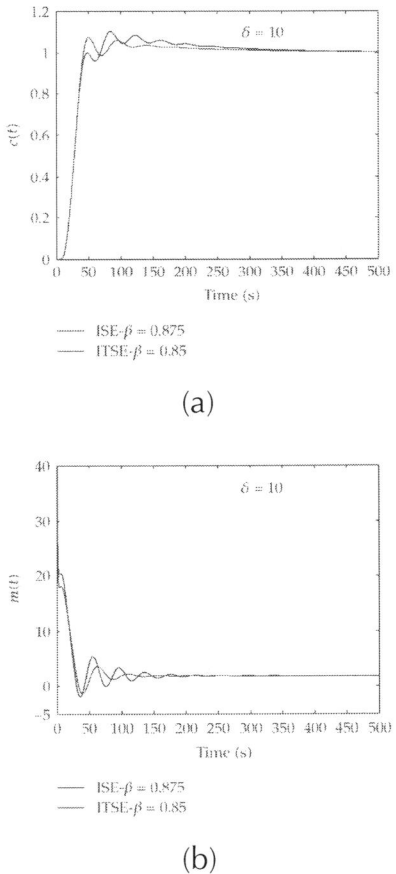

(a)

(b)

Figure 20: Step responses of the closed-loop system and the controller output for the ISE and the ITSE indices, with a PID$^\beta$ controller, $\delta=10$ and x=3.0 m.

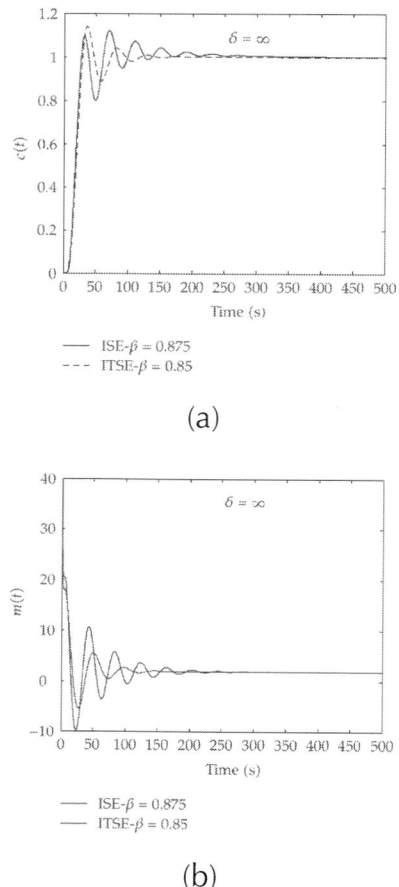

(a)

(b)

Figure 21: Step responses of the closed-loop system and the controller output for the ISE and the ITSE indices, with a PID$^\beta$ controller, $\delta = \infty$ and x=3.0m.

The step responses reveal a large diminishing of the overshoot and the rise time when compared with the integer PID, showing a good transient response and a zero steady-state error. The PID$^\beta$ leads to better results than the classical PID controller tuned through the ZNOL rule. These results demonstrate the effectiveness of the fractional algorithms when used for the control of fractional-order systems. The step response and the controller output are also improved when the saturation level δ is diminished.

Figure 22 depicts the ISE and ITSE indices for $0 \leq \beta \leq 1$, when $\delta = \{20, ..., 100\}$ and $\delta = \infty$. We verify the existence of a minimum for $\beta = 0.875$ and $\beta = 0.85$ for the ISE and ITSE cases, respectively. Furthermore, the higher the δ the lower the value of the index.

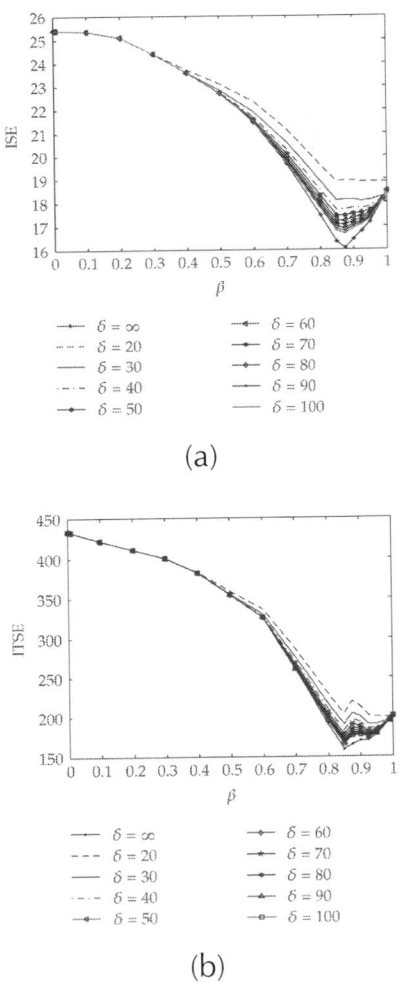

(a)

(b)

Figure 22: ISE and ITSE versus $0 \leq \beta \leq 1$ for $\delta = \{20, ..., 100\}$ and $\delta = \infty$.

Figures 23 and 24 show the variation of the settling time t_s, the peak time t_p, the rise time t_r, and the percent overshoot ov(%), for the closed-loop response tuned through the minimization of the ISE and the ITSE indices, respectively.

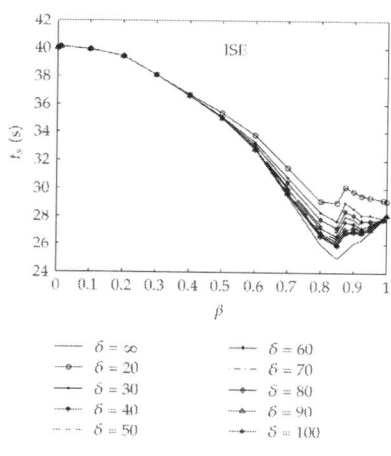

(a)

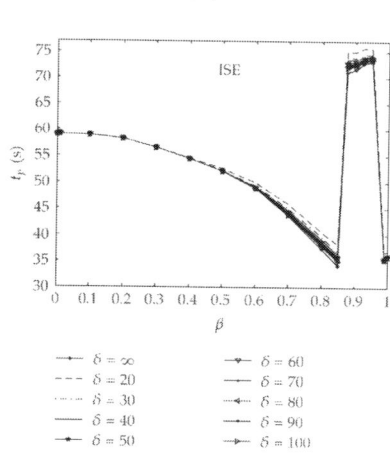

(b)

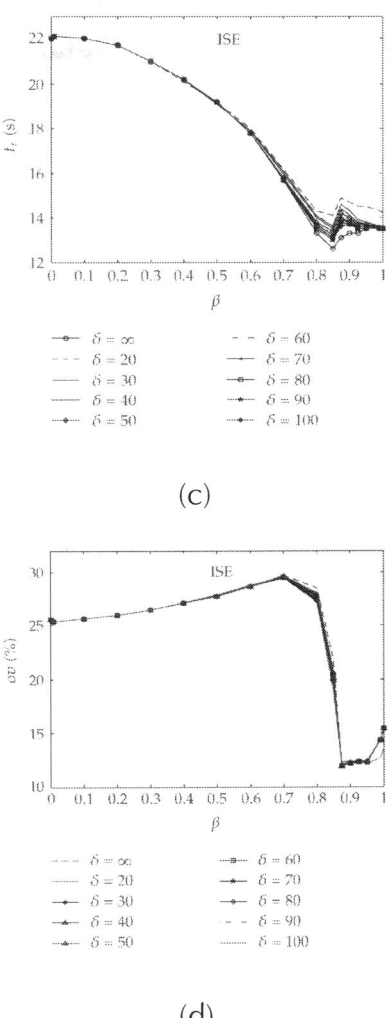

(c)

(d)

Figure 23: Parameters t_s, t_p, t_r, ov(%) for the step responses of the closed-loop system for the ISE indice, with a PID$^\beta$ controller, when $\delta = \{20,....,100\}$ and $\delta = \infty$, x=3.0 m.

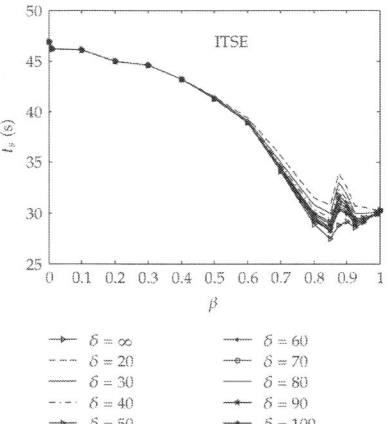

(a)

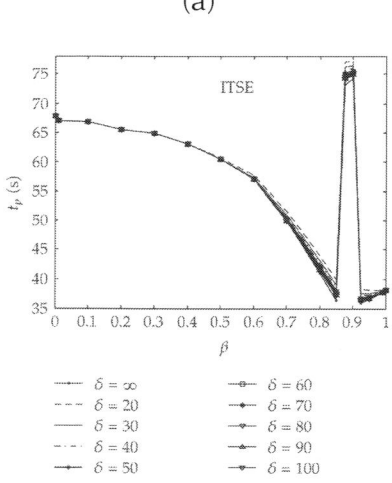

(b)

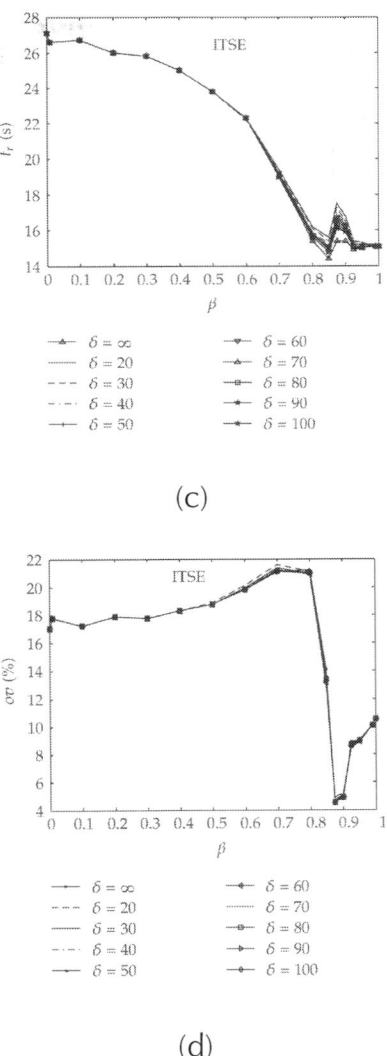

(c)

(d)

Figure 24: Parameters t_s, t_p, t_r, ov(%) for the step responses of the closed-loop system for the ITSE indice, with a PID$^\beta$ controller, when $\delta=\{20,....,100\}$ and $\delta=\infty$, x=3.0 m.

In the ISE case t_s, t_p, and t_r diminish rapidly for $0 \leq \beta \leq 0.875$, while for $\beta > 0.875$ the parameters increase smoothly. For the ITSE, we verify the same behavior for $\beta = 0.85$. On the other hand, $ov(\%)$ increases smoothly for $0 \leq \beta \leq 0.7$, while for $\beta > 0.7$ it decreases very quickly, both for the ISE and the ITSE indices.

In conclusion, for $0.85 \leq \beta \leq 0.875$ we get the best controller tuning, superior to the performance revealed by the classical integer-order scheme.

CIRCUIT SYNTHESIS USING EVOLUTIONARY ALGORITHMS

In recent decades evolutionary computation (EC) techniques have been applied to the design of electronic circuits and systems, leading to a novel area of research called Evolutionary Electronics (EE) or Evolvable Hardware (EH). EE considers the concept for automatic design of electronic systems. Instead of using human conceived models, abstractions, and techniques, EE employs search algorithms to develop implementations not achievable with the traditional design schemes, such as the Karnaugh or the Quine-McCluskey Boolean methods.

Several papers proposed designing combinational logic circuits using evolutionary algorithms and, in particular, genetic algorithms (GAs) [32, 33] and hybrid schemes such as the memetic algorithms (MAs) [34].

Particle swarm optimization (PSO) constitutes an alternative evolutionary computation technique, and this paper studies its application to combinational logic circuit synthesis. Bearing these ideas in mind, the organization of this section is as follows. Section 6.1 presents a brief overview of the PSO. Section 6.2 describes the PSO-based circuit design, while Section 6.3 exhibits the simulation results.

Particle Swarm Optimization

In literature about PSO the term 'swarm intelligence' appears rather often and, therefore, we begin by explaining why this is so.

Noncomputer scientists (ornithologists, biologists, and psychologists) did early research, which led into the theory of particle swarms. In these areas, the term "swarm intelligence" is well known and characterizes the case when a large number of individuals are able of accomplish complex tasks. Motivated by these facts, some basic simulations of swarms were abstracted into the mathematical field. The usage of swarms for solving simple tasks in nature became an intriguing idea in algorithmic and function optimization.

Eberhart and Kennedy were the first to introduce the PSO algorithm [35], which is an optimization method inspired in the collective intelligence of swarms of biological populations, and was discovered through simplified social model simulation of bird flocking, fishing schooling, and swarm theory.

In the PSO, instead of using genetic operators, as in the case of GAs, each particle (individual) adjusts its flying according with its own and its companions experiences. Each particle is treated as a point in a D-dimensional space and is manipulated as described in what follows in the original PSO algorithm:

$$v_{id} = v_{id} + c_1 \text{rand}()(p_{id} - x_{id}) + c_2 \text{Rand}()(p_{gd} - x_{id}), \qquad (6.1a)$$

$$x_{id} = x_{id} + v_{id}, \qquad (6.1b)$$

where c_1 and c_2 are positive constants, $\text{rand}()$ and $\text{Rand}()$ are two random functions in the range $[0,1]$, $X_i = (x_{i1}, x_{i2}, \dots x_{iD})$ represents the ith particle, $P_i = (p_{i1}, p_{i2}, \dots, p_{iD})$ is the best previous position (the position giving the best fitness value) of the particle, the symbol g represents the index of the best particle among all particles in the population, and $V_i=(v_{i1},v_{i2},.....,v_{iD})$ is the rate of the position change (velocity) for particle i.

However, (6.1a) and (6.1b) represent the flying trajectory of a population of particles. Also, (6.1a) describes how the velocity is dynamically updated and (6.1b) the position update of the "flying" particles. Moreover, (6.1b) is divided in three parts, namely the momentum, the cognitive and the social parts. In the first part the velocity cannot be changed abruptly: it is adjusted based on the current velocity. The second part represents the learning from its own flying experience. The third part consists on the learning group flying experience [36].

The first new parameter added into the original PSO algorithm is the inertia weigh. The dynamic equation of PSO with inertia weigh is modified to be

$$v_{id} = wv_{id} + c_1\text{rand}()(p_{id} - x_{id}) + c_2\text{Rand}()(p_{gd} - x_{id}), \tag{6.2a}$$

$$x_{id} = x_{id} + v_{id}, \tag{6.2b}$$

where w constitutes the inertia weigh that introduces a balance between the global and the local search abilities. A large inertia weigh facilitates a global search while a small inertia weigh facilitates a local search.

Another parameter, called constriction coefficient k. is introduced with the hope that it can insure a PSO to converge. A simplified method of incorporating it appears in (6.3), where k is function of c_1 and c_2 as it is presented as follows:

$$v_{id} = k[v_{id} + c_1\text{rand}()(p_{id} - x_{id}) + c_2\text{Rand}()(p_{gd} - x_{id})],$$
$$x_{id} = x_{id} + v_{id}, \tag{6.3}$$

$$k = 2\left(2 - \phi - \sqrt{\phi^2 - 4\phi}\right)^{-1}, \tag{6.4}$$

where $\phi = c_1 + c_2, \phi > 4$.

There are two different PSO topologies, namely, the global version and the local version. In the global version of PSO, each particle flies through the search space with a velocity that is dynamically adjusted according to the particle's personal best performance achieved so far and the best performance achieved so far by all particles. On the other hand, in the local version of PSO, each particle's velocity is adjusted according to its personal best and the best performance achieved so far within its neighborhood. The neighborhood of each particle is generally defined as topologically nearest particles to the particle at each side.

PSO is an evolutionary algorithm simple in concept, easy to implement and computationally efficient. Figures 25, 26, and 27 present a generic EC algorithm, a hybrid algorithm, more precisely a MA and the original procedure for implementing the PSO algorithm, respectively.

```
1. Initialize the population
2. Calculate the fitness of each individual in the
   population
3. Reproduce selected individuals to form a new
   population
4. Perform evolutionary operations such as
   crossover and mutation on the population
5. Loop to step 2 until some condition is met
```

Figure 25: Evolutionary computation algorithm.

```
1. Initialize the population
2. Calculate the fitness of each individual in the
   population
3. Reproduce selected individuals to form a new
   population
4. Perform evolutionary operations such as
   crossover and mutation on the population
5. Apply a local search algorithm
5. Loop to step 2 until some condition is met
```

Figure 26: Memetic algorithm.

1. Initialize population in hyperspace
2. Evaluate fitness of individual particles
3. Modify velocities based on previous best and global (or neighborhood) best
4. Terminate on some condition
5. Go to step 2

Figure 27: Particle swarm optimization process.

The different versions of the PSO algorithms are the real-value PSO, which is the original version of PSO and is well suited for solving real-value problems; the binary version of PSO, which is designed to solve binary problems; and the discrete version of PSO, which is good for solving the event-based problems. To extend the real-value version of PSO to binary/discrete space, the most critical part is to understand the meaning of concepts such as trajectory and velocity in the binary/discrete space.

Kennedy and Eberhart [35] use velocity as a probability to determine whether x_{id} (a bit) will be in one state or another (zero or one). The particle swarm formula of (6.1a) remains unchanged, except that now p_{id} and x_{id} are integers in $[0.0, 1.0]$ and a logistic transformation $S(v_{id})$ is used to accomplish this modification. The resulting change in position is defined by the following rule:

$$\text{if } [\text{rand}() < S(v_{id})] \text{ then } x_{id} = 1; \text{ else } x_{id} = 0, \tag{6.5}$$

where the function $S(v)$ is a sigmoid limiting transformation and rand() is a random number selected from a uniform distribution in the range $[0.0, 1.0]$.

PSO Based Circuit Design

We adopt a PSO algorithm to design combinational logic circuits. A truth table specifies the circuits and the goal is to implement a functional circuit with the least possible complexity. Four sets of logic gates have been defined, as shown in Table 2, being Gset 2 the simplest one (i.e., a RISC-like set) and Gset 6 the most complex gate set (i.e., a CISC-like set). Logic gate named WIRE means a logical no-operation.

Table 2: Gate sets

Gate set	Logic gates
Gset 2	{AND, XOR, WIRE}
Gset 3	{AND, OR, XOR, WIRE}
Gset 4	{AND, OR, XOR, NOT, WIRE}
Gset 6	{AND, OR, XOR, NOT, NAND, NOR, WIRE}

In the PSO scheme the circuits are encoded as a rectangular matrix A (row $\times$ column=r$\times$c) of logic cells as represented in Figure 28.

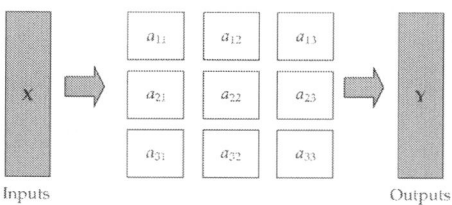

Inputs   Outputs

Figure 28: A 3×3 matrix representing a circuit with input X and output Y.

Three genes represent each cell: <input1><input2><gate type>, where input1 and input2 are one of the circuit inputs, if they are in the first column, or one of the previous outputs, if they are in other columns. The gate type is one of the elements adopted in the gate set. The chromosome is formed with as many triplets as the matrix size demands (e.g., triplets =3×r×c). For example, the chromosome that represents a 3×3 matrix is depicted in Figure 29.

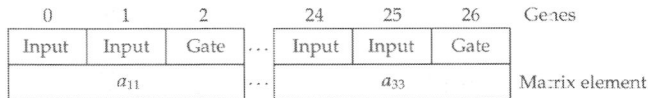

Figure 29: Chromosome for the 3×3 matrix of Figure 28.

The initial population of circuits (particles) has a random generation. The initial velocity of each particle is initialized with zero. The following velocities are calculated applying (6.2a) and the new positions result from using (6.2b). This way, each potential solution, called particle, flies through the problem space. For each gene is calculated the corresponding velocity. Therefore, the new positions are as many as the number of genes in the chromosome. If the new values of the input genes result out of range, then a re-insertion function is used. If the calculated gate gene is not allowed a new valid one is generated at random. These particles then have memory and each keeps information of its previous best position (pbest) and its corresponding fitness. The swarm has the pbest of all the particles and the particle with the greatest fitness is called the global best (gbest).

The basic concept of the PSO technique lies in accelerating each particle towards its pbest and gbest locations with a random weighted acceleration. However, in our case we also use a kind of mutation operator that introduces a new cell in 10% of the population. This mutation operator changes the characteristics of a given cell in the matrix. Therefore, the mutation modifies the gate type and the two inputs, meaning that a completely new cell can appear in the chromosome.

To run the PSO we have also to define the number P of individuals to create the initial population of particles. This population is always the same size across the generations, until reaching the solution.

The calculation of the fitness function F_s in (6.6) has two parts, f_1 and f_2, where f_1 measures the functionality and f_2 measures the simplicity. In a first phase, we compare the output Y produced by the PSO-generated circuit with the required values Y_R, according with the truth table, on a bit-per-bit basis. By other words, f_1 is incremented by one for each correct bit of the output until f_1 reaches the maximum value f_{10} that occurs when we have a functional circuit. Once the circuit is functional, in a second phase, the algorithm tries to generate circuits with the least number of gates. This means that the resulting circuit must have as much genes <gate type> ≡ <wire> as possible. Therefore, the index f_2, that measures the simplicity (the number of null operations), is increased by one (zero) for each wire (gate) of the generated circuit, yielding

$$f_{10} = 2^{ni} \times no,$$

$$f_1 = f_1 + 1 \text{ if } \{\text{bit } i \text{ of } \mathbf{Y}\} = \{\text{bit } i \text{ of } \mathbf{Y_R}\}, \quad i = 1, \ldots, f_{10},$$

$$f_2 = f_2 + 1 \text{ if } gate\ type = wire,$$

$$F_s = \begin{cases} f_1, & F_s < f_{10}, \\ f_1 + f_2, & F_s \geq f_{10}, \end{cases} \tag{6.6}$$

where ni and no represent the number of inputs and outputs of the circuit.

The concept of dynamic fitness function F_d results from an analogy between control systems and the GA case, where we master the population through the fitness function. The simplest control system is the proportional algorithm; nevertheless, there can be other control algorithms, such as the proportional and the differential scheme.

In this line of thought, (6.6) is a static fitness function F_s and corresponds to using a simple proportional algorithm. Therefore, to implement a proportional-derivative evolution the fitness function needs a scheme of the type [18]

$$F_d = F_s + KD^\mu[F_s], \tag{6.7}$$

where $0 \leq \mu \leq 1$ is the differential fractional-order and $K \in \mathbb{R}$ is the "gain" of the dynamical term.

Experiments and Results

A reliable execution and analysis of an EC algorithm usually requires a large number of simulations to provide a reasonable assurance that the stochastic effects are properly considered. Therefore, in this study are developed n=20 simulations for each case under analysis.

The experiments consist on running the three algorithms {GA,MA,PSO} to generate a typical combinational logic circuit, namely, a 2-to-1 multiplexer (M2-1), a 1-bit full adder (FA1), a 4-bit parity checker (PC4) and a 2-bit multiplier (MUL2), using the fitness scheme described in (6.6) and (6.7). The circuits are generated with the gate sets presented in Table 2 and P=3000, w=0.5, c_1=1.5, and c_2=2.

Figure 30 depicts the standard deviation of the number of generations to achieve the solution S(N) versus the average number of generations to achieve the solution Av(N) for the algorithms {GA,MA,PSO}, the circuits {M2-1,FA1,PC4,MUL2}, and the gate sets {2,3,4,6,}. In these figure, we can see that the MUL2circuit is the most complex one, while the PC4 and the M2-1 are the simplest circuits. It is also possible to conclude that Gset 6 is the less efficient gate set for all algorithms and circuits.

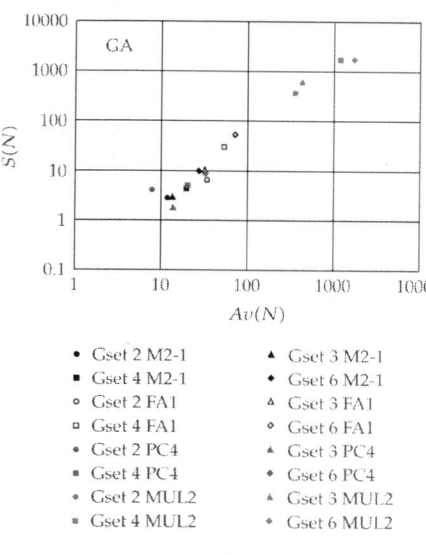

• Gset 2 M2-1	▲ Gset 3 M2-1
■ Gset 4 M2-1	◆ Gset 6 M2-1
○ Gset 2 FA1	▵ Gset 3 FA1
□ Gset 4 FA1	◇ Gset 6 FA1
• Gset 2 PC4	▲ Gset 3 PC4
■ Gset 4 PC4	◆ Gset 6 PC4
• Gset 2 MUL2	▴ Gset 3 MUL2
✳ Gset 4 MUL2	◆ Gset 6 MUL2

(a)

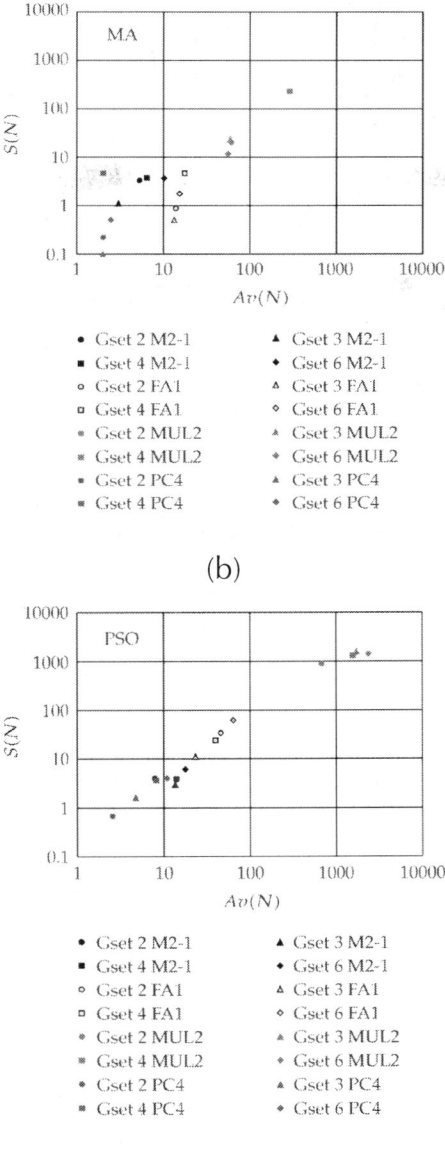

(b)

(c)

Figure 30: S(N) versus Av(N) with P=3000 and F$_s$ for the GA, the MA, and the PSO algorithms.

Figure 30 reveals that the plots follow a power law:

$$S(N) = a[Av(N)]^b \quad a, b \in \mathfrak{R}. \tag{6.8}$$

Table 3 presents the numerical values of the parameters (a,b) for the three algorithms.

Table 3: The parameters (a,b) and (c,d)

Algorithm	a	b	c	d
GA	0.0365	1.602	0.1526	1.1734
MA	0.0728	1.2602	0.2089	1.3587
PSO	0.2677	1.1528	0.0141	1.1233

In terms of S(N) versus Av(N), the MA algorithm presents the best results for all circuits and gate sets. In what concerns the other two algorithms, the PSO is superior (inferior) to the GA for complex (simple) circuits.

Figure 31 depicts the average processing time to obtain the solution Av(PT) versus the average number of generations to achieve the solution Av(N) for the algorithms {GA,MA,PSO}, the circuits {M2-1,FA1,PC4,MUL2} and the gate sets {2,3,4,6}. When analysing these charts it is clear that the PSO algorithm demonstrates to be around ten times faster than the MA and the GA algorithms.

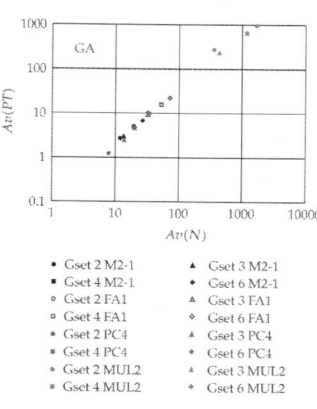

(a)

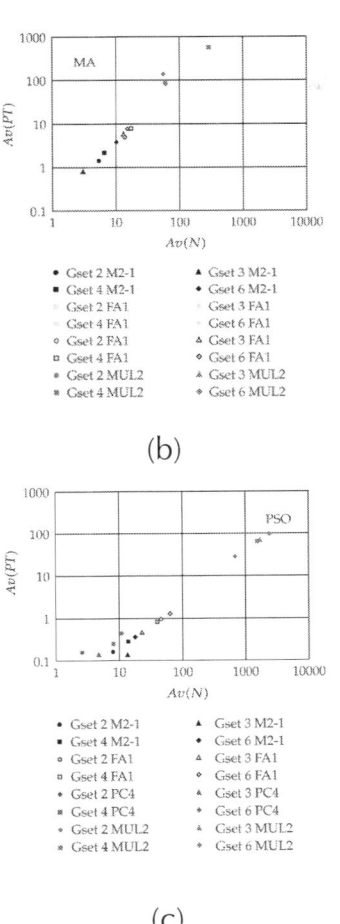

(b)

(c)

Figure 31: Av(PT) versus Av(N) with P=3000 and F_s for the GA, the MA, and the PSO algorithms.

These plots follow also a power law:

$$Av(PT) = c[Av(N)]^d \quad c, d \in \Re. \tag{6.9}$$

Table 3 shows parameters (c,d) and we can see that the PSO algorithm has the best values.

Figures 32 and 33 depict the standard deviation of the number of generations to achieve the solution S(N) and the average processing time to obtain the solution Av(PT), respectively, versus the average number of generations to achieve the solution Av(N) for the PSO algorithm using F_d, the circuits {M2-1, FA1, PC4, MUL2}, and the gate sets {2,3,4,6}. We conclude that F_d leads to better results in particular for the MUL-2circuit and for the Av(PT).

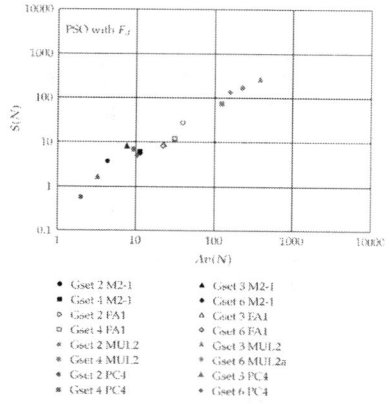

Figure 32: S(N) versus Av(N) for the PSO algorithm, P=3000 and F_d.

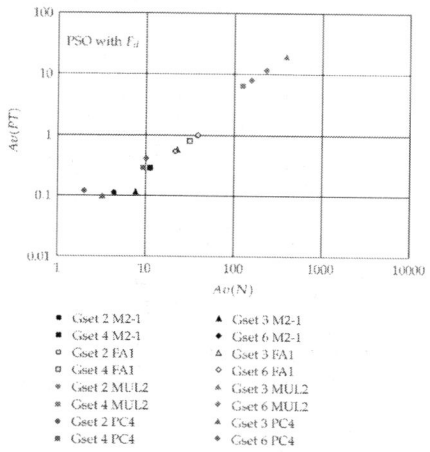

Figure 33: Av(PT) versus Av(N) for the GA, P=3000 and F_d.

Figures 34 and 35 present a comparison between F_s and F_d.

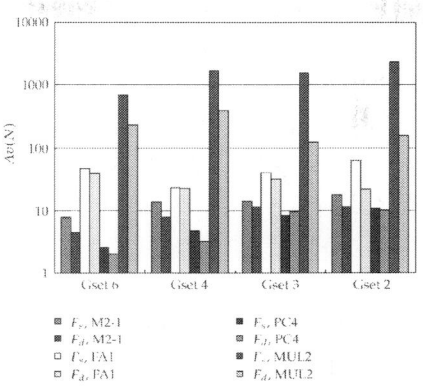

Figure 34: Av(N) for the PSO algorithm, P=3000 using F_s and F_d.

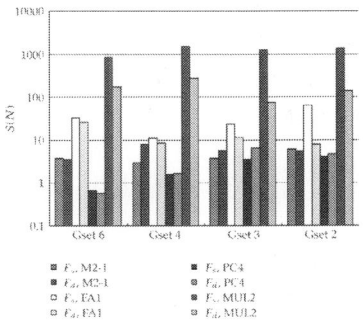

Figure 35: S(N)for the PSO algorithm, P=3000 using F_s and F_d.

In terms of S(N) versus Av(N) it is possible to say that the MA algorithm presents the best results. Nevertheless, when analysing Figure 31, that shows Av(PT) versus Av(N) for reaching the solutions, we verify that the PSO algorithm is very efficient, in particular, for the more complex circuits.

The PSO-based algorithm for the design of combinational circuits follows the same profile as the other two evolutionary techniques presented in this paper.

Adopting the study of the S(N) versus Av(N) for the three evolutionary algorithms, the MA algorithm presents better results over the GA and the PSO algorithms. However, in what concerns the processing time to achieve the solutions, the PSO outcomes clearly the GA and the MA algorithms. Moreover, applying the F_d the results obtained are improved further in all gate sets and in particular for the more complex circuits.

CONCLUSIONS

Fractional Calculus (FC) goes back to the beginning of the theory of differential calculus. Nevertheless, the application of FC just emerged in the last two decades, due to the progress in the area of chaos that revealed subtle relationships with the FC concepts.

Recently FC has been a fruitful field of research in science and engineering and many scientific areas are currently paying wider attention to the FC concepts. In the field of dynamical systems theory, some work has been carried out but the proposed models and algorithms are still in a preliminary stage of establishment. This article presented several case studies on the implementation of FC-based models and control systems, being demonstrated the advantages of using the FC theory in different areas of science and engineering. In fact, this paper studied a variety of different physical systems, namely

i.   tuning of PID controllers using fractional calculus concepts;

ii. fractional PD^α control of a hexapod robot;

iii. fractional dynamics in the trajectory control of redundant manipulators;

iv. heat diffusion;

v.   circuit synthesis using evolutionary algorithms.

It has been recognized the advantageous use of this mathematical tool in the modeling and control of these dynamical systems, and the results demonstrate the importance of Fractional Calculus and motivate for the development of new applications.

REFERENCES

1. K. B. Oldham and J. Spanier, The Fractional Calculus: Theory and Applications of Differentiation and Integration to Arbitrary Order, vol. 11 of Mathematics in Science and Engineering, Academic Press, New York, NY, USA, 1974.

2. K. S. Miller and B. Ross, An Introduction to the Fractional Calculus and Fractional Differential Equations, A Wiley-Interscience Publication, John Wiley & Sons, New York, NY, USA, 1993.

3. I. Podlubny, Fractional Differential Equations, vol. 198 of Mathematics in Science and Engineering, Academic Press, San Diego, Calif, USA, 1999.

4. R. Hilfer, Ed., Applications of Fractional Calculus in Physics, World Scientific Publishing, Singapore, 2000.

5. A. Oustaloup, La Commande CRONE: Commande Robuste d›Ordre Non Entier, Editions Hermès, Paris, France, 1991.

6. A. Oustaloup, La Dérivation Non Entière: Théorie, Synthèse et Applications, Editions Hermès, Paris, France, 1995.

7. R. S. Barbosa, J. A. T. Machado, and I. M. Ferreira, "PID controller tuning using fractional calculus concepts," Fractional Calculus & Applied Analysis, vol. 7, no. 2, pp. 119–134, 2004.

8. R. S. Barbosa, J. A. T. Machado, and I. M. Ferreira, "Tuning of PID controllers based on bode›s ideal transfer function," Nonlinear Dynamics, vol. 38, no. 1–4, pp. 305–321, 2004.

9. M. F. Silva, J. A. T. Machado, and A. M. Lopes, "Comparison of fractional and integer order control of an hexapod robot," in Proceedings of International Design Engineering Technical Conferences and Computers and Information in Engineering Conference, vol. 5 of 19th Biennial Conference on Mechanical Vibration and Noise, pp. 667–676, ASME, Chicago, Ill, USA, September 2003.

10. M. F. Silva, J. A. T. Machado, and I. S. Jesus, "Modelling and simulation of walking robots with 3 dof legs," in Proceedings of the 25th IASTED International Conference on Modelling, Identification and Control (MIC ‹06), pp. 271–276, Lanzarote, Spain, 2006.

11. M. F. Silva, J. A. T. Machado, and A. M. Lopes, "Position/force control of a walking robot," Machine Intelligence and Robot Control, vol. 5, pp. 33–44, 2003.

12. M. F. Silva and J. A. T. Machado, "Fractional order PDα joint control of legged robots," Journal of Vibration and Control, vol. 12, no. 12, pp. 1483–1501, 2006.

13. F. Duarte and J. A. T. Machado, "Chaotic phenomena and fractional-order dynamics in the trajectory control of redundant manipulators," Nonlinear Dynamics, vol. 29, no. 1–4, pp. 315–342, 2002.

14. J. A. T. Machado, "Analysis and design of fractional-order digital control systems," Systems Analysis Modelling Simulation, vol. 27, no. 2-3, pp 107–122, 1997.

15. J. A. T. Machado, "Discrete-time fractional-order controllers," Fractional Calculus & Applied Analysis, vol. 4, no. 1, pp. 47–66, 2001.

16. J. A. T. Machado, I. S. Jesus, J. B. Cunha, and J. K. Tar, "Fractional dynamics and control of distributed parameter systems," Intelligent Systems at the Service of Mankind, vol. 2, pp. 295–305, 2006.

17. I. S. Jesus, R. S. Barbosa, J. A. T. Machado, and J. B. Cunha, "Strategies for the control of heat diffusion systems based on fractional calculus," in Proceedings of the IEEE International Conference on Computational Cybernetics (ICCC ‹06), Budapest, Hungary, 2006.

18. C. Reis, J. A. T. Machado, and J. B. Cunha, "Evolutionary design of combinational circuits using fractional-order fitness," in Proceedings of the 5th Nonlinear Dynamics Conference (EUROMECH ‹05), pp. 1312–1321, 2005.

19. H. W. Bode, Network Analysis and Feedback Amplifier Design, Van Nostrand, New York, NY, USA, 1945.

20. E. S. Conkur and R. Buckingham, "Clarifying the definition of redundancy as used in robotics,"Robotica, vol. 15, no. 5, pp. 583–586, 1997.

21. S. Chiaverini, "Singularity-robust task-priority redundancy resolution for real-time kinematic control of robot manipulators," IEEE Transactions on Robotics and Automation, vol. 13, no. 3, pp. 398–410, 1997.

22. C. A. Klein and C. C. Huang, "Review of pseudoinverse control for use with kinematically redundant manipulators," IEEE Transactions on Systems, Man and Cybernetics, vol. 13, no. 2, pp. 245–250, 1983.

23. T. Yoshikawa, Foundations of Robotics: Analysis and Control, MIT Press, Cambridge, Mass, USA, 1988.

24. J. S. Bay, "Geometry and prediction of drift-free trajectories for redundant machines under pseudoinverse control," International Journal of Robotics Research, vol. 11, no. 1, pp. 41–52, 1992.

25. R. G. Roberts and A. A. Maciejewski, "Singularities, stable surfaces, and the repeatable behavior of kinematically redundant manipulators," International Journal of Robotics Research, vol. 13, no. 1, pp. 70–81, 1994.

26. K. L. Doty, C. Melchiorri, and C. Bonivento, "A theory of generalized inverses applied to robotics,"International Journal of Robotics Research, vol. 12, no. 1, pp. 1–19, 1993.

27. Y. Nakamura, Advanced Robotics: Redundancy and Optimization, Addinson-Wesley, New York, NY, USA, 1991.

28. B. Siciliano, "Kinematic control of redundant robot manipulators: a tutorial," Journal of Intelligent and Robotic Systems, vol. 3, no. 3, pp. 201–212, 1990.

29. W. J. Chung, Y. Yorm, and W. K. Chung, "Inverse kinematics of planar redundant manipulators via virtual links with configuration index," Journal of Robotic Systems, vol. 11, no. 2, pp. 117–128, 1994.

30. S. Seereeram and J. T. Wen, "A global approach to path planning for redundant manipulators," IEEE Transactions on Robotics and Automation, vol. 11, no. 1, pp. 152–159, 1995.

31. M. da Graça Marcos, F. B. M. Duarte, and J. A. T. Machado, "Complex dynamics in the trajectory control of redundant manipulators," Nonlinear Science and Complexity, pp. 134–143, 2007.

32. S. J. Louis, G. J. E. Rawlins, and G. J. Designer, "Genetic algorithms: genetic algorithms in structure design," in Proceedings of the 4th International Conference on Genetic Algorithms, 1991.

33. D. E. Goldberg, Genetic Algorithms in Search Optimization and Machine Learning, Addison-Wesley, New York, NY, USA, 1989.

34. C. Reis, J. A. T. Machado, and J. B. Cunha, "An evolutionary hybrid approach in the design of combinational digital circuits," WSEAS Transactions on Systems, vol. 4, no. 12, pp. 2338–2345, 2005.

35. J. Kennedy and R. C. Eberhart, "Particle swarm optimization," in Proceedings of the IEEE International Conference on Neural Networks, pp. 1942–1948, November 1995.

36. Y. Shi and R. C. Eberhart, "A modified particle swarm optimizer," in Proceedings of the International Conference on Evolutionary Computation, pp. 69–73, May 1998.

CITATION

J. A. Tenreiro Machado, Manuel F. Silva, Ramiro S. Barbosa, et al., "Some Applications of Fractional Calculus in Engineering," Mathematical Problems in Engineering, vol. 2010, Article ID 639801, 34 pages, 2010. doi:10.1155/2010/639801.

Univalence Conditions for Two General Integral Operators

Adriana Oprea[1] and Daniel Breaz[2]

[1]Department of Mathematics, University of Pitești, Pitești, Romania

[2]Department of Mathematics, "1 Decembrie 1918" University of Alba Iulia, Alba Iulia, Romania

ABSTRACT

Let A be the class of all analytic functions which are analytic in the open unit disc $U = \{z : |z| < 1\}$. In this paper we study the problem of univalence for the following general integral operators:

$$F_n(z) = \int_0^z \prod_{i=1}^n \left(\frac{f_i(t)}{t} e^{g_i(t)} \right)^{\alpha_i} dt,$$

$G_n(z) = \int_0^z \prod_{i=1}^n (f_i'(t) e^{g_i(t)})^{\beta_i} dt$, in the open unit disc U, when f_i, $g_i \in A$, α_i, $\beta_i \in C$.

INTRODUCTION

Let $U = \{z : |z| < 1\}$ be the unit disk and A be the class of all functions of the form

$$f(z) = z + \sum_{k=2}^{\infty} a_k z^k, \quad z \in U \tag{1}$$

Which are analytic in U and satisfy the conditions

$$f(0) = f'(0) - 1 = 0.$$

We denote by S the class of univalent and regular functions.

In order to derive our main results, we have to recall here the following univalence conditions.

Theorem 1.1:

[1] (Becker's univalence criterion).

If the function f is regular in unit disk U, $f(z) = z + a_2 z^2 + \cdots$ and

$$\left(1 - |z|^2\right)\left|\frac{z f''(z)}{f'(z)}\right| \leq 1, \text{ for all } z \in U \qquad (2)$$

Then the function f is univalent in U.

Theorem 1.2:

[2] If the function g is regular in U and $|g(z)| < 1$ in U, then for all $\xi \in U$ the following inequalities hold

$$\left|\frac{g(\xi) - g(z)}{1 - \overline{g(z)} g(\xi)}\right| \leq \left|\frac{\xi - z}{1 - \overline{z}\xi}\right| \qquad (3)$$

And

$$|g'(z)| \leq \frac{1 - |g(z)|^2}{1 - |z|^2}.$$

The equalities hold in case $g(z) = \varepsilon \dfrac{z+u}{1+\bar{u}z}$ where $|\varepsilon| = 1$ and $|u| < 1$.

Remark 1.3:

[2] For $z = 0$, from inequality (3) we obtain for every $\xi \in U$

$$\left| \frac{g(\xi) - g(0)}{1 - \overline{g(0)}g(\xi)} \right| \leq |\xi| \tag{4}$$

And, hence

$$|g(\xi)| \leq \frac{|\xi| + |g(0)|}{1 + |g(0)||\xi|} \tag{5}$$

Considering $g(0) = a$ and $\xi \in z$, then

$$|g(z)| \leq \frac{|z| + |a|}{1 + |a||z|},$$

For all $z \in U$.

MAIN RESULTS

In this paper we study the univalence of the following general integral operators:

$$F_n(z) = \int_0^z \prod_{i=1}^n \left(\frac{f_i(t)}{t} e^{g_i(t)} \right)^{\alpha_i} dt, \tag{6}$$

Where $f_i, g_i \in A$ and $\alpha_i \in C$,

$$G_n(z) = \int_0^z \prod_{i=1}^n \left(f_i'(t) e^{g_i(t)} \right)^{\beta_i} dt, \tag{7}$$

Where $f_i, g_i \in A$ and $\beta_i \in C$.

Theorem 2.1:

Let $\alpha_n \in C$, $f_n \in S$, $f_n(z) = z + a_2^n z^2 + \cdots$, $n \in N^*$, $g_n \in S$, $g_n(z) = z + b_2^n z^2 + \cdots$, $n \in N^*$ If

$$\left| \frac{z f_n'(z) - f_n(z)}{z f_n(z)} \right| \leq 1, \tag{8}$$

For all $n \in N^*$, for all $z \in U$ and

$$\left| g_n'(z) \right| \leq 1$$

$$\frac{\left| \alpha_1 \right| + \left| \alpha_2 \right| + \cdots + \left| \alpha_n \right|}{\left| \alpha_1 \alpha_2 \cdots \alpha_n \right|} < 1, \tag{9}$$

$$\left| \alpha_1 \alpha_2 \cdots \alpha_n \right| \leq \frac{1}{\max\limits_{|z| \leq 1} \left[2 \left(1 - |z|^2 \right) |z| \dfrac{|z| + |c|}{1 + |c||z|} \right]}. \tag{10}$$

Where

$$|c| = \frac{\left| \alpha_1 \left(a_2^1 + 1 \right) + \cdots + \alpha_n \left(a_2^n + 1 \right) \right|}{2 \left| \alpha_1 \alpha_2 \cdots \alpha_n \right|}$$

Then the function

$$F_n(z) = \int_0^z \prod_{i=1}^n \left(\frac{f_i(t)}{t} e^{g_i(t)} \right)^{\alpha_i} dt, \qquad (11)$$

Is in the class S.

Proof:

We have $f_n \in S$, $\dfrac{f_n(z)}{z} \neq 0$, for all $n \in N^*$ and

$$\left(\frac{f_1(z)}{z} e^{g_1(z)} \right)^{\alpha_1} \cdots \left(\frac{f_n(z)}{z} e^{g_n(z)} \right)^{\alpha_n} = 1,$$

When $z = 0$.

Let us consider the function:

$$h(z) = \frac{1}{2|\alpha_1 \alpha_2 \cdots \alpha_n|} \frac{F_n''(z)}{F_n'(z)}. \qquad (12)$$

From (6), we have:

$$F_n'(z) = \prod_{i=1}^n \left(\frac{f_i(z)}{z} e^{g_i(z)} \right)^{\alpha_i} \qquad (13)$$

and

$$F_n''(z) = \sum_{i=1}^n \alpha_i \left(\frac{f_i(z)}{z} e^{g_i(z)} \right)^{\alpha_i - 1} \left(\frac{z f_i'(z) - f_i(z)}{z^2} e^{g_i(z)} + \frac{f_i(z)}{z} e^{g_i(z)} g_i'(z) \right) \prod_{\substack{k=1 \\ k \neq i}}^n \left(\frac{f_k(z)}{z} e^{g_k(z)} \right)^{\alpha_k}. \qquad (14)$$

From (13) and (14), we have:

$$\frac{F_n''(z)}{F_n'(z)} = \sum_{i=1}^{n} \alpha_i \left(\frac{zf_i'(z) - f_i(z)}{zf_i(z)} + g_i'(z) \right).$$

Using relations before the function h has the form:

$$h(z) = \frac{1}{2|\alpha_1\alpha_2\cdots\alpha_n|} \sum_{i=1}^{n} \alpha_i \left(\frac{zf_i'(z) - f_i(z)}{zf_i(z)} + g_i'(z) \right).$$

(15)

We have:

$$h(0) = \frac{1}{2|\alpha_1\alpha_2\cdots\alpha_n|} \alpha_1 \left(a_2^1 + 1 \right) + \frac{1}{2|\alpha_1\alpha_2\cdots\alpha_n|} \alpha_2 \left(a_2^2 + 1 \right) + \cdots + \frac{1}{2|\alpha_1\alpha_2\cdots\alpha_n|} \alpha_n \left(a_2^n + 1 \right).$$

By using the relations (15), (8) and (9), we obtain:

$$|h(z)| \leq \frac{1}{2|\alpha_1\alpha_2\cdots\alpha_n|} \sum_{i=1}^{n} \left| \alpha_i \left(\frac{zf_i'(z) - f_i(z)}{zf_i(z)} + g_i'(z) \right) \right| \leq \frac{1}{2|\alpha_1\alpha_2\cdots\alpha_n|} 2\sum_{i=1}^{n} |\alpha_i| \leq 1$$

(16)

$$|h(0)| = \frac{\left| \alpha_1 \left(a_2^1 + 1 \right) + \cdots + \alpha_n \left(a_2^n + 1 \right) \right|}{2|\alpha_1\alpha_2\cdots\alpha_n|} = |c|.$$

(17)

Applying Remark 1.3 for the function h, we obtain:

$$|h(z)| = \frac{1}{2|\alpha_1\alpha_2\cdots\alpha_n|} \left| \frac{F_n''(z)}{F_n'(z)} \right| \leq \frac{|z| + |h(0)|}{1 + |h(0)||z|} \leq \frac{|z| + |c|}{1 + |c||z|}.$$

(18)

From (18), we get:

$$\left|\left(1-|z|^2\right)z\frac{F_n''(z)}{F_n'(z)}\right|\leq|\alpha_1\alpha_2\cdots\alpha_n|2\left(1-|z|^2\right)|z|\frac{|z|+|c|}{1+|c||z|},$$

(19)

For all $z \in U$.

Let us consider the function: $H:[0,1] \to R$

$$H(x)=2\left(1-x^2\right)x\frac{x+|c|}{1+|c|x}, \quad x=|z|.$$

Since $H\left(\dfrac{1}{2}\right)=\dfrac{3}{4}\dfrac{1+2|c|}{2+|c|}>0$, it results:

$$\max_{x\in[0,1]} H(x) > 0.$$

Using this result and the form (19), we have:

$$\left|\left(1-|z|^2\right)z\frac{F_n''(z)}{F_n'(z)}\right|\leq\left|\prod_{i=1}^{n}\alpha_i\right|\max_{|z|<1}\left[2\left(1-|z|^2\right)|z|\frac{|z|+|c|}{1+|c||z|}\right],$$

(20)

For all $z \in U$.

Applying the condition (10) in relation (20), we obtain:

$$\left(1-|z|^2\right)\left|\frac{zF_n''(z)}{F_n'(z)}\right|\leq 1,$$

For all $z \in U$ and from Theorem 1.1, we have $F_n \in S$.

Corollary 2.2:

Let α be a complex number and the functions $f \in S$, $f(z) = z + a_2 z^2 + \cdots$, $g \in S$, $g(z) = z + b_2 z^2 + \cdots$.

If

$$\left| \frac{zf'(z) - f(z)}{zf(z)} \right| < 1 \quad \text{and} \quad \left| g'(z) \right| < 1$$

(21)

For all $z \in U$ and the constant $|\alpha|$ satisfies the condition:

$$|\alpha| \leq \frac{1}{\max\limits_{|z| \leq 1} \left[2|z|\left(1 - |z|^2\right) \dfrac{2|z| + |a_2 + 1|}{2 + |a_2 + 1||z|} \right]},$$

(22)

Then the function

$$F_1(z) = \int_0^z \left(\frac{f(t)}{t} e^{g(t)} \right)^\alpha \, dt,$$

(23)

Is in the class S.

Proof:

We consider $n = 1$ in Theorem 2.1.

Remark 2.3:

For $n = 1$, $e^{g_1(t)} = 1$, $\alpha_1 = 1$ and $f_1 = f$ in relation (11), we obtain the integral operator $I(z) = \int_0^z \frac{f(t)}{t} dt$, introduced by J. W. Alexander in [3] .

Remark 2.4:

For $n = 1$, $e^{g_1(t)} = 1$, $\alpha_1 = \alpha$, $f_1 = f$ in relation (6), we obtain the integral operator $F(z) = \int_0^z \left(\frac{f(t)}{t} \right)^\alpha dt$, defined and studied by V. Pescar in [4] [5] .

Remark 2.5:

For $e^{g_i(t)} = 1$, for all $i = 1, \cdots, n$, we get the integral operator $I_n(z) = \int_0^1 \prod_{i=1}^n \left(\frac{f(t)}{t} \right)^{\alpha_i} dt$, $z \in U$ studied by D. Breaz, N. Breaz in [6] and D. Breaz in [7].

Theorem 2.6

Let $\beta_n \in C$, $f_n \in S$, $f_n(z) = z + a_2^n z^2 + \cdots$, $n \in N^*$, $g_n \in S$, $g_n(z) = z + b_2^n z^2 + \cdots$, $n \in N^*$.

If

$$\left| \frac{f_n''(z)}{f_n'(z)} \right| \leq 1,$$

(24)

For all $n \in N^*$, for all $z \in U$ and $|g_n'(z)| \leq 1$

$$\frac{|\beta_1| + |\beta_2| + \cdots + |\beta_n|}{|\beta_1 \beta_2 \cdots \beta_n|} < 1,$$

(25)

$$\left| \prod_{i=1}^n \beta_i \right| \leq \frac{1}{\max\limits_{|z| \leq 1} \left[2 \left(1 - |z|^2 \right) |z| \frac{|z| + |c|}{1 + |c||z|} \right]},$$

(26)

Where

$$|c| = \frac{\left| \beta_1 \left(2a_2^1 + 1 \right) + \cdots + \beta_n \left(2a_2^n + 1 \right) \right|}{2 |\beta_1 \beta_2 \cdots \beta_n|}$$

Then the function

$$G_n(z) = \int_0^z \prod_{i=1}^n \left(f_i'(t) e^{g_i(t)} \right)^{\beta_i} dt,$$

(27)

Is in the class S.

$$\left(f_1'(z) e^{g_1(z)} \right)^{\beta_1} \cdots \left(f_n'(z) e^{g_n(z)} \right)^{\beta_n} = 1,$$

/

When $z = 0$.

Let us consider the function:

$$p(z) = \frac{1}{2|\beta_1\beta_2 \cdots \beta_n|} \frac{G_n''(z)}{G_n'(z)}.$$

(28)

From (27), we have:

$$G_n'(z) = \prod_{i=1}^n \left(f_i'(z) e^{g_i(z)} \right)^{\beta_i}$$

(29)

and

$$G_n''(z) = \sum_{i=1}^n \beta_i \left(f_i'(z) e^{g_i(z)} \right)^{\beta_i - 1} \left(f_i''(z) e^{g_i(z)} + f_i'(z) e^{g_i(z)} g_i'(z) \right) \prod_{\substack{k=1 \\ k \neq i}}^n \left(f_k'(z) e^{g_k(z)} \right)^{\beta_k}.$$

(30)

From (29) and (30), we get:

$$\frac{G_n''(z)}{G_n'(z)} = \sum_{i=1}^n \beta_i \left(\frac{f_i''(z)}{f_i'(z)} + g_i'(z) \right).$$

(31)

Using relation (31) the function p has the form:

$$p(z) = \frac{1}{2|\beta_1\beta_2\cdots\beta_n|} \sum_{i=1}^{n} \beta_i \left(\frac{f_i''(z)}{f_i'(z)} + g_i'(z) \right).$$

We have:

$$p(0) = \frac{\beta_1\left(2a_2^1+1\right) + \beta_2\left(2a_2^2+1\right) + \cdots + \beta_n\left(2a_2^n+1\right)}{2|\beta_1\beta_2\cdots\beta_n|}.$$

By using the relations (24), (25) and (28), we obtain:

$$|p(z)| \le \frac{1}{2|\beta_1\beta_2\cdots\beta_n|} \sum_{i=1}^{n} \left| \beta_i \left(\frac{f_i''(z)}{f_i'(z)} + g_i'(z) \right) \right| \le \frac{1}{2|\beta_1\beta_2\cdots\beta_n|} 2\sum_{i=1}^{n} |\beta_i| \le 1 \tag{32}$$

and

$$|p(0)| = \frac{\left|\beta_1\left(2a_2^1+1\right) + \beta_2\left(2a_2^2+1\right) + \cdots + \beta_n\left(2a_2^n+1\right)\right|}{2|\beta_1\beta_2\cdots\beta_n|} = |c|. \tag{33}$$

Applying Remark 1.3 for the function p, we obtain:

$$|p(z)| = \frac{1}{2|\beta_1\beta_2\cdots\beta_n|} \left| \frac{G''(z)}{G'(z)} \right| \le \frac{|z| + |p(0)|}{1 + |p(0)||z|} \le \frac{|z| + |c|}{1 + |c||z|}. \tag{34}$$

From (34), we get:

$$\left| \left(1 - |z|^2\right) z \frac{G_n''(z)}{G_n'(z)} \right| \le |\beta_1\beta_2\cdots\beta_n| 2\left(1 - |z|^2\right)|z|\frac{|z| + |c|}{1 + |c||z|}, \tag{35}$$

for all $z \in U$.

Let us consider the function $Q:[0,1] \to R$

$$Q(x) = 2\left(1-x^2\right)x\frac{x+|c|}{1+|c|x}, \quad x = |z|.$$

Since $Q\left(\dfrac{1}{2}\right) = \dfrac{3}{4}\dfrac{1+2|c|}{2+|c|} > 0$, it results:

$$\max_{x \in [0,1]} Q(x) > 0.$$

Using this result and the form (35), we have:

$$\left|\left(1-|z|^2\right)z\frac{G_n''(z)}{G_n'(z)}\right| \le \left|\prod_{i=1}^{n}\beta_i\right|\max_{|z|<1}\left[2\left(1-|z|^2\right)|z|\frac{|z|+|c|}{1+|c||z|}\right],$$

(36)

for all $z \in U$.

Applying the condition (26) in relation (36), we obtain:

$$\left(1-|z|^2\right)\left|\frac{zF_n''(z)}{F_n'(z)}\right| \le 1,$$

for all $z \in U$ and from Theorem 1.1, we have $G_n \in S$.

Corollary 2.7:

Let β be a complex number and the functions $f \in S$, $f(z) = z + a_2 z^2 + \cdots$, $g \in S$, $g(z) = z + b_2 z^2 + \cdots$.
If

$$\left|\frac{f''(z)}{f'(z)}\right| < 1 \quad \text{and} \quad \left|g'(z)\right| < 1$$

(37)

for all $z \in U$ and the constant $|\beta|$ satisfies the condition:

$$|\beta| \le \frac{1}{\max\limits_{|z| \le 1} \left[2|z|\left(1-|z|^2\right) \frac{2|z|+|2a_2+1|}{2+|2a_2+1||z|} \right]},$$

(38)

Then the function

$$G_1(z) = \int_0^z \left(f'(t) e^{g(t)} \right)^\beta \, dt,$$

(39)

Is in the class S.

Proof:

We consider $n = 1$ in Theorem 2.6.

Remark 2.8:

For $n = 1$, $e^{g_1(t)} = 1$, $\beta_1 = \beta$, $f_1 = f$ in relation (27), we obtain the integral operator $G_\beta(z) = \int_0^z (f'(t))^\beta dt$, defined and studied by V. Pescar in [8] [9].

Remark 2.9:

For $n = 1$ and $\beta = \alpha$ in relation (27), we obtain the integral operator $I_1(f,g)(z) = \leftarrow \int_0^z (f'(t)e^{g(t)})^\alpha dt$, introduced and studied by N. Ularu and D. Breaz in [10] and [11].

ACKNOWLEDGMENTS

This work was supported by the strategic project PERFORM, POSDRU 159/1.5/S/138963, inside POSDRU Romania 2014, co-financed by the European Social Fund-Investing in People.

REFERENCES

1. Becker, J. (1972) Lownersche Differentialgleichung und quasikonform fortsetz-bare schlichte Funktionen. Journal für die Reine und Angewandte Mathematik, 255, 23-43.
2. Nehari, Z. (1952) Conformal Mapping. McGraw-Hill Book Company, New York.
3. Alexander, J.W. (1915) Functions Which Map the Interior of the Unit Circle upon Simple Regions. Annals of Mathematics, 17, 12-22. http://dx.doi.org/10.2307/2007212
4. Pescar, V. (1997) On Some Integral Operations Which Preserve the Univalence. Journal of Mathematics, 30, 1-10.
5. Pescar, V. (1998) On the Univalence of an Integral Operator. Studia Universitatis "Babes-Bolyai", Cluj-Napoca, Mathematica, 43, 95-97.
6. Breaz, D. and Breaz, N. (2002) Two Integral Operators. Studia Universitatis "Babes-Bolyai", Cluj-Napoca, Mathematica, 3, 13-21.
7. Breaz, D. (2008) Certain Integral Operators on the Classes $M(\beta i)$ and $N(\beta i)$. Journal of Inequalities and Applications, Article ID: 719354.
8. Pescar, V. (1997) Some Integral Operators and Their Univalence. The Journal of Analysis, 5, 157-162.
9. Pescar, V. (1997) An Integral Operator Which Preserves the Univalency. The Annual Conference of the Romanian Society of Mathematical Sciences, Bucharest, 29 May-1 June 1997, 179-181.
10. Ularu, N. and Breaz, D. (2012) Univalence Criterion and Convexity for an Integral Operator. Applied Mathematics Letter, 25, 658-661.http://dx.doi.org/10.1016/j.aml.2011.10.011
11. Ularu, N. and Breaz, D. (2013) Univalence Condition and Properties for Two Integral Operators. Applied Sciences, 15, 112-117.

CITATION

Oprea, A. and Breaz, D. (2014) Univalence Conditions for Two General Integral Operators. Advances in Pure Mathematics, 4, 487-493. doi: 10.4236/apm.2014.48054.

Integral Inequalities of Gronwall-Bellman Type

Zareen A. Khan
Department of Mathematics, Princess Noura Bint
Abdurehman University, Riyadh, KSA

9

ABSTRACT

The goal of the present paper is to establish some new approach on the basic integral inequality of Gronwall-Bellman type and its generalizations involving function of one independent variable which provides explicit bounds on unknown functions. The inequalities given here can be used as tools in the qualitative theory of certain partial differential and integral equations.

INTRODUCTION

The Gronwall type integral inequalities provide a necessary tool for the study of the theory of differential equations, integral equations and inequalities of the various types. Some applications of this result can be used to the study of existence, uniqueness theory of differential equations and the stability of the solution of linear and nonlinear differential equations. During the past few years, several authors have established several Gronwall type integral inequalities in one or two independent real variables [1] -[15] . Of course, such results have application in the theory of partial differential equations and Volterra integral equations.

Closely related to the foregoing first-order ordinary differential operators is the following result of Bellman [11] : If the functions g(t) and u(t) are nonnegative for $t \geq 0$, and if $C \geq 0$, the inequality

$$u(t) \leq c + \int_0^t g(s)u(s)ds, \quad t \geq 0$$

implies that

$$u(t) \leq c \exp\left(\int_0^t g(s)ds\right), \quad t \geq 0$$

(1.1)

Our aim in this paper is to establish new explicit bounds on some basic integral inequalities of one independent variable which will be equally important in handling the inequality (1.1). Given application in this paper is also illustrating the usefulness of our result.

MAIN RESULTS

Lemma 2.1:

Let u(t) and g(t) be nonnegative continuous functions defined for $I = [0,\infty)$. Let $k(t) > 1$ defined for $I = [0,\infty)$ and also k'(t) be nonnegative continuous functions defined for $I = [0,\infty)$. If

$$u(t) \leq k(t) + \int_0^t g(s)u(s)ds, \quad \forall t \in I$$

(2.1)

Then

$$u(t) \leq k(0)\exp\left(k(t) - k(0) + \int_0^t g(s)ds\right), \quad \forall t \in I$$

(2.2)

Proof:

Define a function m(t) by the right-hand side of (2.1), such that

$$m(t) = k(t) + \int_0^t g(s)u(s)\,ds$$

(2.3)

where

$$m(0) = k(0)$$

(2.4)

Then m(t)>1. From (2.1) and (2.3), we observe that

$$u(t) \le m(t)$$

(2.5)

Differentiating both sides of (2.3) with respect to t, we get

$$m'(t) = k'(t) + g(t)u(t)$$

By using (2.5) and since k(t)>1, the above equation can be restated as

$$\frac{m'(t)}{m(t)} \le k'(t) + g(t)$$

(2.6)

Integrating both sides of (2.6) from 0 to t and also using (2.4), we observe that

$$m(t) \le k(0)\exp\left(k(t) - k(0) + \int_0^t g(s)\,ds \right)$$

(2.7)

From (2.5) and (2.7), we get the required inequality (2.2).

Theorem 2.2:

Let u(t), f(t)and g(t) be nonnegative continuous functions defined for $I = [0, \infty)$. Let $k(t) > 1$ defined for $I = [0, \infty)$ and also k'(t) be nonnegative continuous functions defined for $I = [0, \infty)$. If

$$u(t) \leq k(t) + \int_0^t f(s)u(s)ds + \int_0^t f(s)\left(\int_0^s g(\partial)u(\partial)d\partial\right)ds, \quad \forall t \in I$$

(2.8)

Then

$$u(t) \leq k(t) + k(0)\int_0^t f(s)\exp\left(k(s) - k(0) + \int_0^s (f(\partial) + g(\partial))d\partial\right)ds, \quad \forall t \in I$$

(2.9)

Proof:

Define a function m(t) by the right-hand side of (2.8), such that

$$m(t) = k(t) + \int_0^t f(s)u(s)ds + \int_0^t f(s)\left(\int_0^s g(\partial)u(\partial)d\partial\right)ds, \quad \forall t \in I$$

(2.10)

where

$$m(0) = k(0)$$

(2.11)

Then m(t)>1. From (2.9) and (2.10), we observe that

$$u(t) \leq m(t)$$

(2.12)

Differentiating both sides of (2.10) with respect to t, we get

$$m'(t) = k'(t) + f(t)\left[u(t) + \int_0^t g(s)u(s)ds\right]$$

By using (2.12), the above equation can be restated as

$$m'(t) \le k'(t) + f(t)v(t)$$

(2.13)

where

$$v(t) = m(t) + \int_0^t g(s)m(s)ds$$

(2.14)

and

$$v(0) = m(0) = k(0)$$

(2.15)

Again differentiating both sides of (2.14) with respect to x and using (2.13) and using the fact that m(t) $m(t) \le v(t)$, we get

$$\frac{v'(t)}{v(t)} \le k'(t) + \left[f(t) + g(t) \right]$$

(2.16)

By applying Lemma 2.1 implies the estimation of v(t) as

$$v(t) \le k(0)\exp\left(k(t) - k(0) + \int_0^t \left[f(0) + g(0) \right]ds \right)$$

(2.17)

By substituting (2.17) in (2.13), we have

$$m'(t) \le k'(t) + k(0)f(t)\exp\left(k(t) - k(0) + \int_0^t \left[f(s) + g(s) \right]ds \right)$$

Integrating both sides of the above inequality from 0 to t and also using (2.11), we observe that

$$m(t) \le k(t) + k(0) \int_0^t f(s) \exp\left(k(s) - k(0) + \int_0^s [f(\partial) + g(\partial)] d\partial \right) ds \tag{2.18}$$

From (2.12) and (2.18), we get the required inequality (2.9). This completes the proof.

Theorem 2.3:

Let u(t), f(t)and g(t),k(t), and k'(t) be defined as in Theorem 2.2. If

$$u(t) \le k(t) + \int_0^t f(s)u(s)ds + \int_0^t f(s)\left(\int_0^s f(\tau)\left(\int_0^\tau g(\partial)u(\partial)d\partial \right) d\tau \right) ds, \quad \forall t \in I \tag{2.19}$$

Then

$$u(t) \le k(t) + \int_0^t f(s)\left[k(s) + k(0)\int_0^s f(\tau)\exp\left(k(\tau) - k(0) + \int_0^\tau (f(\partial) + g(\partial))d\partial \right) d\tau \right] ds, \quad \forall t \in I$$

Proof:

The proof of Theorem 2.3 is the same as the proof of Theorem 2.2 and by applying the Lemma 2.1 with suitable modifications.

APPLICATION

As an application, let us consider the bound for the solution of Volterra integral equation of the form

$$x(t) = f(t) + \int_0^t p(t,s)g(t,x(s),Tx(s))ds \tag{3.1}$$

where x, f and g are the elements of R^n, p(t,s) is a $n \times n$ matrix, $g \in C[I \times R^n \times R^n, R^n]$ and $x \in C[I, R^n]$ and T be a continuous operator such that T maps C(I) into C(I).

Define

$$\left| p(t,s) \right| \le 1 \tag{3.2}$$

and

$$\left| g(t, x, y) \right| \le f(t) \left[|x| + |y| \right], \quad t \in I \tag{3.3}$$

Also let $\left| f(t) \right| \le k(t)$, where $k(t) > 1$ $\tag{3.4}$

$$\left| Tx(t) \right| \le \int_0^t g(s) \left| x(s) \right| ds, \quad t \in I \tag{3.5}$$

Then

$$\left| x(t) \right| \le k(t) + k(0) \int_0^t f(s) \exp\left(k(s) - k(0) + \int_0^s \left(f(\partial) + g(\partial) \right) d\partial \right) ds, \quad \forall t \in I$$

Proof:

Taking absolute value of the both sides of (3.1), we get

$$\left| x(t) \right| \le \left| f(t) \right| + \int_0^t \left| p(t,s) g(s, x(s), Tx(s)) \right| ds \tag{3.6}$$

By substituting from (3.2), (3.3), (3.4) and (3.5) in (3.6), we have

$$\left|x(t)\right| \le k(t) + \int_0^t f(s)\left|x(s)\right|ds + \int_0^t f(s)\left(\int_0^s g(\partial)\left|x(\partial)\right|d\partial\right)ds, \quad \forall t \in I$$

The remaining proof will be the same as the proof of Theorem 2.2 with suitable modifications. We note that Theorem 2.2 can be used to study the stability, boundedness and continuous dependence of the solutions of (3.1).

CONCLUSIONS

We finally mention that the integral inequalities obtained in this paper allow us to study the stability, boundedness and asymptotic behavior of the solutions of a class of more general partial differential and integral equations.

REFERENCES

1. Abdeldaim, A. and Yakout, M. (2011) On Some New Integral Inequalities of Gronwall-Bellman-Pachpatte Type. Applied Mathematics and Computation, 217, 7887-7899.http://dx.doi.org/10.1016/j.amc.2011.02.093
2. Pachpatte, B.G. (2001) On Some Fundamental Integral Inequalities and Their Discrete Analogues. Journal of Inequalities in Pure and Applied Mathematics, 2, Article 15.
3. Pachpatte, B.G. (1994) On Some Fundamental Integral Inequalities Arising in the Theory of Differential Equations. Chinese Journal of Mathematics, 22, 261-273.
4. Pachpatte, B.G. (1996) Comparison Theorems Related to a Certain Inequality Used in the Theory of Differential Equations. Soochow Journal of Mathematics, 22, 383-394.
5. Langenhop, C.E. (1960) Bounds on the Norm of a Solution of a General Differential Equation. Proceedings of the American Mathematical Society, 11, 795-799.
6. Bainov, D. and Simeonov, P. (1992) Integral Inequalities and Applications. Kluwer Academic Publishers, Dordrecht.

7. Mitrinovíc, D.S., Pĕcaríc, J.E. and Fink, A.M. (1991) Inequalities Involving Functions and Their Integrals and Derivatives. Kluwer Academic Publishers, Dordrecht.

8. Beckenbach, E.F. and Bellman, R. (1961) Inequalities. Springer-Verlag, New York.http://dx.doi.org/10.1007/978-3-642-64971-4

9. Bihari, I. (1956) A Generalization of a Lemma of Bellman and Its Application to Uniqueness Problem of Differential Equations. Acta Mathematica Academiae Scientiarum Hungarica, 7, 71-94. http://dx.doi.org/10.1007/BF02022967

10. Guiliano, L. (1946) Generalazzioni di un lemma di Gronwall. Rend. Accad., Lincei, 1264-1271.

11. Bellman, R. (1943) The Stability of Solutions of Linear Differential Equations. Duke Mathematical Journal, 10, 643- 647. http://dx.doi.org/10.1215/S0012-7094-43-01059-2

12. Dragomir, S.S. and Kim, Y.H. (2002) On Certain New Integral Inequalities and Their Applications. JIPAM, 3, Article 65.

13. Dragomir, S.S. and Kim, Y.H. (2003) Some Integral Inequalities for Functions of Two Variables. Electronic Journal of Differential Equations, 10, 1-13.

14. Gronwall, T.H. (1919) Note on the Derivatives with Respect to a Parameter of Solutions of a System of Differential Equations. Annals of Mathematics, 20, 292-296.http://dx.doi.org/10.2307/1967124

15. Nemyckii, V.V. and Stepanov, V.V. (1947) Qualitative Theory of Differential Equations. OGIZ, Moscow.

CATITON

Khan, Z. (2014) Integral Inequalities of Gronwall-Bellman Type. Applied Mathematics, 5, 3484-3488. doi: 10.4236/am.2014.521326.

Computation of the Multivariate Normal Integral over a Complex Subspace

Kartlos Joseph Kachiashvili[1, 2] and Muntazim Abbas Hashmi[3]

[1]Abdul Salam School of Mathematical Sciences, GC University, Lahore, Pakistan [2]Vekua Institute of Applied Mathematics, Tbilisi State University, Tbilisi, Georgia
[3]Air University Multan Campus, Multan, Pakistan

ABSTRACT

The computation of the multivariate normal integral over a Complex Subspace is a challenge, especially when the integration region is of a complex nature. Such integrals are met with, for example, in the generalized Neyman-Pearson criterion, conditional Bayesian problems of testing many hypotheses and so on. The Monte-Carlo methods could be used for their computation, but at increasing dimensionality of the integral the computation time increases unjustifiedly. Therefore a method of computation of such integrals by series after reduction of dimensionality to one without information loss is offered below. The calculation results are given.

INTRODUCTION

At testing many hypotheses with reference to the parameters of multivariate normal distribution, the problem of computation of multivariate normal integrals over a Complex Subspace of the following form arises [1]

$$p_{ij} = \int_{\Gamma_j} p(\mathbf{x}|H_i)\,d\mathbf{x},\, i,j = 1,\cdots,S, i \neq j, \tag{1}$$

where S is the number of tested hypotheses $H_i : \theta = \theta^i$, supposing that sample $x^T = (x_1, \ldots, x_n)$ was brought about by distribution

$$p\left(\mathbf{x}, \boldsymbol{\theta}^i\right) = p\left(x_1, \cdots, x_n; \theta_1^i, \cdots, \theta_m^i\right) \equiv p\left(\mathbf{x} \mid H_i\right), i = 1, \cdots, S$$

where $\theta^T = (\theta_1, \ldots, \theta_m)$ is the vector of distribution parameters and Γ_i is the acceptance region of hypothesis H_i from sample space $(x \in R^n)$ R^n, which has the following form

$$\Gamma_j = \left\{ \mathbf{x} : k_j^j p\left(\mathbf{x} \mid H_j\right) > \sum_{\ell=1, \ell \neq j}^{S} k_\ell^j p\left(\mathbf{x} \mid H_\ell\right) \right\},$$

$$j = 1, \cdots, S$$

(2)

Where

$$0 \leq k_\ell^j < +\infty, \quad \ell = 1, \cdots, S .$$

Such regions of hypotheses acceptance arise, for example, in the generalized Neyman-Pearson criterion, and also in conditional Bayesian problems of testing many hypotheses [2,3]. The dimensionality of these integrals often reaches several tens when practical problems are solved. For example, in ecological problems the number of controlled parameters, according to which the decision is made, is quite often equal to several tens [4]; in the air defence problems, in particular, in the problems of tracking of flying objects using radar measurement information, the dimensionality of the problem is equal to the multiplication of the number of flying objects by the number of surveys made by the radar set [5] and so on. On the other hand, the time for solution of these problems is often limited and at times it plays a decisive role especially at solving the defence problems.

It is known that the complexity of realization and the obtained accuracy of numerical methods of computation of multidimensional integrals depend heavily on the dimensionality of these integrals and the complexity of the integration region configuration. In the considered case the integration regions are nonconvex and quite complex. Therefore it is difficult to realize the numerical methods and to provide the desired accuracy of calculation even when the dimensionality of integral is greater than or equal to three [6]. The methods of computation of the multivariate normal integral on the hyperrectangle offered in [7-12] are unsuitable for this case because of the complexity of the integration region.

Despite the convenience and the simplicity of computations, the Monte Carlo method is computer time consuming, especially at large dimensionality of integrals [3, 13, 14]. Therefore the method of approximate computation of integral (1) for a very short period of time is topical in many applications of mathematical statistics [15, 16].

The aim of the present paper is the development of the method of computation of probability integral (1) with the desired accuracy in a minimum of time.

PROBLEM STATEMENT

Let us consider the case when the probability distribution density of the vector x looks like

$$p(\mathbf{x}|H_i)$$

$$= (2\pi)^{-n/2} |W_i|^{-1/2} \exp\left\{-\frac{1}{2}(\mathbf{x}-\mathbf{a}^i)^T \mathbf{W}_i^{-1}(\mathbf{x}-\mathbf{a}^i)\right\},$$

$$i = 1, \cdots, S,$$
\hfill (3)

where

$$\mathbf{a}^{iT} = \left(a_1^i, \cdots, a_n^i\right), \quad \mathbf{W}_i = \begin{pmatrix} \sigma_1^{i2} & \rho_{12}^i & \cdots & \rho_{1n}^i \\ \rho_{21}^i & \sigma_2^{i2} & \cdots & \rho_{2n}^i \\ \cdots & \cdots & \cdots & \cdots \\ \rho_{n1}^i & \rho_{n2}^i & \cdots & \sigma_n^{i2} \end{pmatrix}$$

For probability distribution density (3), let us rewrite decision-making region (2) as

$$\Gamma_j = \left\{ \mathbf{x} : \sum_{\ell=1}^{S} C_\ell^j \exp(-y_\ell) < 0 \right\},$$

$$(4)$$

Where

$$C_\ell^j = k_\ell^j (2\pi)^{-n/2} \left| \mathbf{W}_\ell \right|^{-1/2}, \ell \neq j,$$

$$C_j^j = -k_j^j (2\pi)^{-n/2} \left| \mathbf{W}_j \right|^{-1/2},$$

$$y_\ell = \frac{1}{2} (\mathbf{x} - \mathbf{a}^\ell)^T \mathbf{W}_\ell^{-1} (\mathbf{x} - \mathbf{a}^\ell), \ell, j = 1, \cdots, S.$$

$$(5)$$

Random variables $y_\ell, \ell = 1,...,S$, are squared forms of the normally distributed random vector, and, if hypothesis H_i is true, their mathematical expectations are equal to

$$E(y_\ell | H_i) = \frac{1}{2} (\mathbf{a}^i - \mathbf{a}^\ell)^T \mathbf{W}_\ell^{-1} (\mathbf{a}^i - \mathbf{a}^\ell)$$

$$+ \frac{1}{2} \operatorname{trace} (\mathbf{W}_i \mathbf{W}_\ell^{-1}), \quad \ell, i = 1, \cdots, S.$$

$$(6)$$

Therefore, if hypothesis H_i is true, the random variable y_ℓ has noncentral distribution x^2 with the degree of freedom n and with the parameter of noncentrality equal to (6) [2,17,18].

It is obvious that, at $\ell = i$ and hypothesis H_i is true, the random variable y_i has the central x^2 distribution with the degree of freedom n.

Let us write down (1) as follows

$$P_{ij} = \int_{\Gamma_j} p(\mathbf{x} | H_i) d\mathbf{x} = P \left(\sum_{\ell=1}^{S} C_\ell^j \exp(-y_\ell) < 0 \middle| H_i \right)$$

$$(7)$$

The task consists in the computation of probability (7). The method of its analytical computation is not known so far. For its computation it is possible, for example, to use a modified Monte-Carlo method (with the purpose of reducing the computation time) [3]. Though, at large S, it still takes a good deal of time. The method of computation of probability (7) if hypotheses are formulated with reference only to the mathematical expectation of normally distributed random vector is offered in [3]. This method is unsuitable here, as the random variable

$$\xi_j = \sum_{\ell=1}^{S} C_\ell^j \exp(-y_\ell),$$

(8)

which formulates integration region (4), in [3] is the weighted sum of log-normally distributed random quantities; C_ℓ^j and y_ℓ are determined by formulae (5). In our case, ξ_j is the weighted sum of the exponents of negative quadratic forms of the normally distributed random vector with correlated components.

Let us consider the case S=2. In this case, regions (2) take the form

$$\Gamma_1 = \left\{ \mathbf{x} : p\left(\mathbf{x}|H_2\right) < k_1 p\left(\mathbf{x}|H_1\right) \right\},$$

$$\Gamma_2 = \left\{ \mathbf{x} : p\left(\mathbf{x}|H_1\right) < k_2 p\left(\mathbf{x}|H_2\right) \right\}.$$

With taking into account probability densities (3), for these regions we derive

$$\Gamma_1 = \left\{ \mathbf{x} : \mathbf{x}^T \mathbf{W}_1^{-1} \mathbf{x} - \mathbf{x}^T W_2^{-1} \mathbf{x} + 2\left(\mathbf{a}^{2T} \mathbf{W}_2^{-1} - \mathbf{a}^{1T} \mathbf{W}_1^{-1}\right) \mathbf{x} < \lambda_{12} \right\},$$

$$\Gamma_2 = \left\{ \mathbf{x} : \mathbf{x}^T \mathbf{W}_2^{-1} \mathbf{x} - \mathbf{x}^T W_1^{-1} \mathbf{x} + 2\left(\mathbf{a}^{1T} \mathbf{W}_1^{-1} - \mathbf{a}^{2T} \mathbf{W}_2^{-1}\right) \mathbf{x} < \lambda_{21} \right\},$$

where

$$\lambda_{12} = 2\ln\left(k_1 \frac{|\mathbf{W}_1|^{-\frac{1}{2}}}{|\mathbf{W}_2|^{-\frac{1}{2}}}\right) + \left(\mathbf{a}^{2T}\mathbf{W}_2^{-1}\mathbf{a}^2 - \mathbf{a}^{1T}\mathbf{W}_1^{-1}\mathbf{a}^1\right),$$

$$\lambda_{21} = 2\ln\left(k_2 \frac{|\mathbf{W}_2|^{-\frac{1}{2}}}{|\mathbf{W}_1|^{-\frac{1}{2}}}\right) + \left(\mathbf{a}^{1T}\mathbf{W}_1^{-1}\mathbf{a}^1 - \mathbf{a}^{2T}\mathbf{W}_2^{-1}\mathbf{a}^2\right).$$

Let us designate

$$\xi_{12} = \mathbf{x}^T\mathbf{W}_1^{-1}\mathbf{x} - \mathbf{x}^T\mathbf{W}_2^{-1}\mathbf{x} + 2\left(\mathbf{a}^{2T}\mathbf{W}_2^{-1} - \mathbf{a}^{1T}\mathbf{W}_1^{-1}\right)x,$$

$$\xi_{21} = \mathbf{x}^T\mathbf{W}_2^{-1}\mathbf{x} - \mathbf{x}^T\mathbf{W}_1^{-1}\mathbf{x} + 2\left(\mathbf{a}^{1T}\mathbf{W}_1^{-1} - \mathbf{a}^{2T}\mathbf{W}_2^{-1}\right)x.$$

Then, finally, for the required regions, we shall obtain

$$\Gamma_1 = \left\{\mathbf{x} : \xi_{12} < \lambda_{12}\right\},$$

$$\Gamma_2 = \left\{\mathbf{x} : \xi_{21} < \lambda_{21}\right\}.$$

Each of random variables ξ_{12} and ξ_{21} is the sum of three random variables one of which is distributed by the normal law, and the two others are distributed by the x^2 law. Therefore, the probability distribution laws of random variables ξ_{12} and ξ_{21} have not closed forms.

Thus, at S=2, i.e. at testing two hypotheses with respect to all parameters of multivariate normal distribution (in contradistinction to the case when hypotheses are formulated with respect to only the vector of mathematical expectation [3]), the principal complexity of the considered problem does not decrease.

COMPUTATION OF PROBABILITY INTEGRAL (7) BY SERIES

Let us use the expanded form of representation of the quadratic form in (8) [18,19]. Then

$$\xi_j = \sum_{\ell=1}^{S} C_\ell^j \exp\left\{ \frac{1}{2} \sum_{t_1=1}^{n} \sum_{t_2=1}^{n} \alpha_{t_1,t_2}^\ell \left(\frac{x_{t_1} - a_{t_1}^\ell}{\sigma_{t_1}^\ell} \right) \left(\frac{x_{t_2} - a_{t_2}^\ell}{\sigma_{t_2}^\ell} \right) \right\} \tag{9}$$

where α_{t_1,t_2}^ℓ are the coefficients determined unambiguously by the elements of matrix W_ℓ (see formula (3)).

Let $P_j(z|H_i)$ be the conditional density of probability distribution of the random variable ξ_j. Then, for (7), we obtain

$$p_{ij} = \int_{-\infty}^{0} P_j(z \mid H_i) dz . \tag{10}$$

Here the infinite interval $(-\infty, +\infty)$ is taken as the domain of definition of random variable ξ_j because of the signs of coefficients C_ℓ^j from (5).

As was mentioned above, the probability distribution law of the random variable ξ_j has not a closed form. Let us consider the opportunity of approximating this density by series. For this reason we need the moments of the random variable ξ_j [19-21]. Let us consider the problem of obtaining of these moments.

With this purpose let us calculate the initial moment of the r th order of random variable ξ_j provided that hypothesis H_i is true

$$\mu_r^{j,i} = E\left[\left(\xi_j \right)^r \mid H_i \right] = E\left[\left(\sum_{\ell=1}^{S} C_\ell^j \exp(-y_\ell) \right)^r \mid H_i \right]$$

$$= \sum_{\ell_1=1}^{S} \cdots \sum_{\ell_r=1}^{S} C_{\ell_1}^{j} \cdots C_{\ell_r}^{j} E\left[\exp\left(-\left(y_{\ell_1} + \dots + y_{\ell_r}\right)\right) \middle| H_i \right],$$

$$r = 1, 2, \cdots \tag{11}$$

Expression $y_{e_1} + \dots + y_{e_r}$ is the sum of correlated Quadratic Forms distributed by noncentral x^2 probability distribution laws. Because of correlation, the property of reproducibility of the x^2 distribution does not take place [2, 18], and, consequently the mathematical expectation in (12) has not a closed form.

Let us use power series expansion of the exponent

$$\exp\left(-\left(y_{\ell_1} + \dots + y_{\ell_r}\right)\right) = \sum_{v=0}^{\infty} (-1)^v \frac{1}{v!}\left(y_{\ell_1} + y_{\ell_2} + \dots + y_{\ell_r}\right)^v$$

$$= \sum_{v=0}^{\infty} (-1)^v \frac{1}{v!} \sum_{p_i \in \{\ell_1, \cdots, \ell_r\}; i=1, \cdots, v} y_{p_1} \cdot y_{p_2} \cdots y_{p_v}. \tag{12}$$

Let us use the expanded representation of quadratic form (9) and be satisfied with the first M terms of expansion (12). Then expression for calculation of moments (11) can be represented as follows

$$\mu_r^{j,i} \approx \sum_{\ell_1=1}^{S} \cdots \sum_{\ell_r=1}^{S} C_{\ell_1}^{j} \cdots C_{\ell_r}^{j} \left\{ 1 + \sum_{v=1}^{M} \left(-\frac{1}{2}\right)^v \right.$$

$$\cdot \frac{1}{v!} \left[\sum_{p_i \in \{\ell_1, \cdots, \ell_r\}; i=1, \cdots, v} \sum_{t_1=1}^{n} \sum_{t_2=1}^{n} \alpha_{t_1, t_2}^{p_1} \right.$$

$$\left. \cdots \alpha_{t_1, t_2}^{p_v} E\left[\prod_{\xi=1}^{\Lambda} \left(\frac{x_\xi - a_\xi^{p_\xi}}{\sigma_\xi^{p_\xi}}\right)^{m_\xi} \middle| H_i \right] \right] \right\}, \tag{13}$$

Where

$$\Lambda \in \{1,\cdots,2v\}, \; m_\xi \in \{0,1,\cdots,2v\}$$

and

$$\sum_{\xi=1}^{\Lambda} m_\xi = 2v .$$

Expression (13) contains product moments [17, 19,2 0] of the 2v (v=1,...,M) orders of normalized components of the correlated normally distributed random observation vector. Therefore, they are not equal to zero [18]. A lot of works are dedicated to the problem of computation of product moments [see, for example, 22-28].

In [22] the following problem was solved. Let $x_1, x_2,...,x_n$ be random variables with mutually independent distributions, and let $x = \prod_{i=1}^{n} x_i$. There is found the probability that X lies between A and B, i.e.P{A≤X≤B}, by using the central limit theorem in accordance with which the random variable $\ln X = \prod_{i=1}^{n} \ln x_i$ is approximately distributed by the normal law. The better is this approximation the bigger is n.

The variance of the product of two random variables was studied by Barnett (1955) and Goodman (1960), in the case when they do not need to be independent. Shellard (1952) studied the case when the distribution of $\prod_{i=1}^{n} x_i$ was (approximately) logarithmic-normal. The author considered the case when $x_1, x_2,...,x_n$ are random variables with mutually independent distributions. For finding the probability that $x = \prod_{i=1}^{n} x_i$ lies between A and B, i.e.P{A≤X≤B}, the central limit theorem is used to approach the probability distribution of the random variable ln X by normal distribution and this approach is better at increasing n. In work [25] no assumption is made about the distribution of $\prod_{i=1}^{k} x_i$. There is discussed the case when the K random variables, $x_1,x_2,...,x_k,(K≥2)$ are mutually independent, and the case when they do not have to be independent, and there are obtained their variance formulae. These results are generalizations of the results presented in [24].

In [26] are given exact formulae for the mathematical expectation of $(\overline{x}_i - \overline{X}_i)(\overline{x}_j - \overline{X}_j)(\overline{x}_k - \overline{X}_k)$ and $(\overline{x}_i - \overline{X}_i)(\overline{x}_j - \overline{X}_j)(\overline{x}_k - \overline{X}_k)(\overline{x}_h - \overline{X}_h)$, $(i \neq j \neq k \neq h)$, where $\overline{x}_i$ is the sample mean of the i th "character" in a sample of n elements from a population of N elements and $\overline{X}_i$ is the corresponding population mean. Formulae for estimating these product moments from the sample were also given. These estimations are slightly biased. In [27] the unbiased estimate of the 4-variate product moment was obtained. Asymptotic results for the 3-variate and 4-variate product moments and their estimates were also obtained.

In [28] is derived a formula for the product moment $EX_1^{m_1}...X_p^{m_p}, m_1 \geq 1,...,m_p \geq 1$, in terms of the joint survival function when $(X_1,..., X_p)$ is a non-negative random vector.

From the given review (of course incomplete, because this is not the aim of this paper) of the works dedicated to the study of product moments, it is seen that the problem considered here differs from them.

Theorem 3.1. The initial moment of the r th order of random variable ξ_j determined by (9), provided that hypothesis H_i is true, can be calculated with any specified accuracy by the formula

$$
\mu_r^{j,i} \approx \sum_{\ell_1=1}^{S} \cdots \sum_{\ell_r=1}^{S} C_{\ell_1}^j \cdots C_{\ell_r}^j \left\{ 1 + \sum_{v=1}^{M} \left(-\frac{1}{2} \right)^v \frac{1}{v!} \right.
$$

$$
\times \left[\sum_{p_i \in \{\ell_1,\cdots,\ell_r\}; i=1,\cdots,v} \sum_{t_1=1}^{n} \sum_{t_2=1}^{n} \alpha_{t_1,t_2}^{p_1} \cdots \right.
$$

$$
\alpha_{t_1,t_2}^{p_v} J^{i,\Lambda} \sum_{\tau_i \in \{0,1,2,\cdots,2v\}; i=1,\cdots,\Lambda} d_{\tau_1,\cdots,\tau_\Lambda}^{i,\Lambda}
$$

$$
\left. \left. \times \prod_{\eta=1}^{\Lambda} \sum_{j=0}^{\tau_\eta} \binom{\tau_\eta}{j} \mu_{\tau_\eta - 1}(\mu_1')^j \right] \right\}, r = 1, 2, 3, \cdots
$$

$$
(14)
$$

Where

$$J^{i,\Lambda} = \mathrm{mod}\left| \mathbf{K}^{i,\Lambda} \left(\beta^{i,\Lambda} \right)^{-1} \right|; \quad \beta^{i,2v}$$

and $K^{i,2v}$ are the matrices of eigenvectors and eigenvalues of the inverse covariance matrix of normalized random variables

$$\frac{x_\tau - a_\tau^{p_\tau}}{\sigma_\tau^{p_\tau}}, \tau = 1, \cdots, \Lambda \; ; \quad d_{\tau_1, \cdots, \tau_\Lambda}^{i,\Lambda} \quad , \quad \tau_i \in \{0,1,2,\cdots,2v\},$$

i=1,....,, are the coefficients determined by the terms of matrices $\beta^{i,2v}$ and $K^{i,2v}$ and vector $b^{i,2v}$; μ_1' and $\mu_{\tau_n - j}$ are the initial and central moments of the first and $\tau_n - j$ orders, respectively, defined by formulae (22), (23) and (24).

Proof:

If hypothesis Hi is true, the values $\dfrac{x_\tau - a_\tau^{p_\tau}}{\sigma_\tau^{p_\tau}}$

$\tau = 1, 2, \ldots, \Lambda$, are correlated normally distributed random variables with the parameters

$$E\left(\frac{x_\tau - a_\tau^{p_\tau}}{\sigma_\tau^{p_\tau}} \middle| H_i \right) = \frac{a_\tau^i - a_\tau^{p_\tau}}{\sigma_\tau^{p_\tau}} = b_\tau^{i,p_\tau},$$

$$V\left(\frac{x_\tau - a_\tau^{p_\tau}}{\sigma_\tau^{p_\tau}} \middle| H_i \right) = \frac{\left(\sigma_\tau^i\right)^2}{\left(\sigma_\tau^{p_\tau}\right)^2} = v_\tau^{i,p_\tau}, \tau = 1, \cdots, \Lambda;$$

$$\mathrm{cov}\left[\left(\frac{x_{\tau_1} - a_{\tau_1}^{p_{\tau_1}}}{\sigma_{\tau_1}^{p_{\tau_1}}} \right)\left(\frac{x_{\tau_2} - a_{\tau_2}^{p_{\tau_2}}}{\sigma_{\tau_2}^{p_{\tau_2}}} \right) \middle| H_i \right] = \frac{\rho_{\tau_1, \tau_2}^i}{\sigma_{\tau_1}^{p_{\tau_1}} \cdot \sigma_{\tau_2}^{p_{\tau_2}}}$$

$$= v^{i,p_{\tau_1},p_{\tau_2}}_{\tau_1,\tau_2}, \tau_1, \tau_2 = 1, \cdots, \Lambda. \tag{15}$$

Thus, for calculation of moments (13), it is required to calculate the product moments of -dimensional $(\wedge \in \{1,...,2v\}, v = 1,...,M)$ normally distributed random vectors for which the components of the vectors of mathematical expectations and the covariance matrices are calculated by formulae (15).

Let us designate

$$\mathbf{b}^{i,\Lambda} = \left(b_1^{i,p_1}, \cdots, b_\Lambda^{i,p_\Lambda} \right)^T_{1\times\Lambda},$$

$$\mathbf{V}^{i,\Lambda} = \begin{pmatrix} v_{1,1}^{i,p_1,p_1} & v_{1,2}^{i,p_1,p_2} & \cdots & v_{1,\Lambda}^{i,p_1,p_\Lambda} \\ v_{2,1}^{i,p_2,p_1} & v_{2,2}^{i,p_2,p_2} & \cdots & v_{2,\Lambda}^{i,p_2,p_\Lambda} \\ \cdots & \cdots & \cdots & \cdots \\ v_{\Lambda,1}^{i,p_T,p_1} & v_{\Lambda,2}^{i,p_T,p_2} & \cdots & v_{\Lambda,\Lambda}^{i,p_\Lambda,p_\Lambda} \end{pmatrix}_{\Lambda\times\Lambda} \tag{16}$$

and the corresponding random vector—by

$$\mathbf{y} = \left(y_1, \cdots, y_\Lambda \right)^T, i.e.$$, i.e.

$$y_1 = \frac{x_1 - a_1^{p_1}}{\sigma_1^{p_1}}, \quad y_2 = \frac{x_2 - a_2^{p_2}}{\sigma_2^{p_2}}, \cdots, \quad y_\Lambda = \frac{x_\Lambda - a_\Lambda^{p_\Lambda}}{\sigma_\Lambda^{p_\Lambda}}.$$

For calculation of conditional product moments of the 2v-order, we have

$$E\left(y_1^{m_1} y_2^{m_2} \cdots y_\Lambda^{m_\Lambda} \middle| H_i \right) = \int_{-\infty}^{+\infty} \cdots \int_{-\infty}^{+\infty} y_1^{m_1} y_2^{m_2} \cdots y_\Lambda^{m_\Lambda}$$

$$\times f\left(y_1, y_2, \cdots, y_\Lambda \middle| H_i \right) \mathrm{d}y_1 \mathrm{d}y_2 \cdots \mathrm{d}y_\Lambda, \tag{17}$$

where $f(y_1, y_2, \ldots, y_\Lambda)$ is the Λ-dimensional normal probability distribution density with the vector of mathematical expectations and the covariance matrix calculated by formulae (16).

It is known that the value of integral (17) is invariant to linear transformation of the components of vector x [18] with the accuracy of Jacobian of Transformation. Let us designate the matrixes of eigenvectors and eigenvalues of matrix $(V^{i,\Lambda})^{-1}$ by $\beta^{i,\Lambda} = \left\| \beta^i_{\ell j} \right\|_{\Lambda \times \Lambda}$ and $K^{i,\Lambda} = \left\| K^i_{\ell j} \right\|_{\Lambda \times \Lambda}$, respectively. It should be pointed out that $K^{i,\Lambda}$ is a diagonal matrix. Then the components of Λ- dimensional random vector

$$Z^{i,\Lambda} = \beta^{i,\Lambda} \left(K^{i,\Lambda} \right)^{-1} \left(y - b^{i,\Lambda} \right), \tag{18}$$

will be uncorrelated and will have standard normal distribution of probabilities [3,20].

From (18), we write

$$y = K^{i,\Lambda} \left(\beta^{i,\Lambda} \right)^{-1} Z^{i,\Lambda} + b^{i,\Lambda} .$$

Let us introduce the following designation

$$\gamma^{i,\Lambda} = \left\| \gamma^{i,\Lambda}_{\ell j} \right\|_{\Lambda \times \Lambda} = K^{i,\Lambda} \left(\beta^{i,\Lambda} \right)^{-1} .$$

Then, for the elements of the vector y, we obtain the following expression

$$y_\tau = \sum_{t=1}^{\Lambda} \gamma^{i,\Lambda}_{\tau,t} z^{i,\Lambda}_t + b^{i,\Lambda}_\tau , \tau = 1, \cdots, \Lambda . \tag{19}$$

Using transformation (19), for mathematical expectation (17), we obtain

$$E\left(y_1^{m_1}\,y_2^{m_2}\cdots y_\Lambda^{m_\Lambda}\,\big|\,H_i\right)=\mathbf{J}^{i,\Lambda}$$

$$\times E\left[\prod_{\tau=1}^{\Lambda}\left(\sum_{t=1}^{\Lambda}\gamma_{\tau,t}^{i,\Lambda}z_t^{i,\Lambda}+b_\tau^{i,\Lambda}\right)^{m_\tau}\bigg|\,H_i\right],\tag{20}$$

where $J^{i,\Lambda}=\mathrm{mod}\left|K^{i,\Lambda}\left(\beta^{i,\Lambda}\right)^{-1}\right|$ is the Jacobian of Transformation (18).

Let us raise to the powers the linear forms in the righthand side of expression (20) and group the identical items. Then (20) can be written as

$$E\left(y_1^{m_1}\,y_2^{m_2}\cdots y_\Lambda^{m_\Lambda}\,\big|\,H_i\right)$$

$$=\mathbf{J}^{i,\Lambda}\times\sum_{\tau_i\in\{0,1,2,\cdots,2v\};i=1,\cdots,\Lambda}d_{\tau_1,\cdots,\tau_\Lambda}^{i,\Lambda}\prod_{\eta=1}^{\Lambda}E\left[\left(z_\eta^{i,\Lambda}\right)^{\tau_\eta}\bigg|\,H_i\right],\tag{21}$$

where, the coefficients of the identical items in (20) are designated by $d_{\tau_1,\cdots,\tau_\Lambda}^{i,\Lambda}$, $\tau_i\in\{0,1,2,\ldots,2v\}; i=1,\ldots,\Lambda$; the items of the vector $Z^{i,\Lambda}$ are determined as

$$z_\eta^{i,\Lambda}=\sum_{\ell=1}^{\Lambda}\left[\left(\sum_{\delta=1}^{\Lambda}\beta_{\eta,\delta}^{i}K_{\delta,\ell}^{i}\right)\left(y_\ell-b_\ell^{i,P_\ell}\right)\right],$$

$$\eta=1,\cdots,\Lambda.$$

It is known that [21,29]

$$E\left[\left(z_\eta^{i,\Lambda}\right)^{\tau_\eta}\bigg|\,H_i\right]=\mu_{\tau_\eta}'$$

$$=\sum_{j=0}^{\tau_\eta}\binom{\tau_\eta}{j}\mu_{\tau_\eta-j}\left(\mu_1'\right)^j,\tag{22}$$

where μ_{τ_η}' and $\mu_{\tau_\eta-j}$ are the initial and central moments of τ_η and $\tau_\eta-j$ orders, respectively, of random variable $z_\eta^{i,\Lambda}$. After simple routine transformations, for the considered case we obtain

$$\mu_j = \begin{cases} \dfrac{\left(\vartheta_{z,\eta}^{i,\Lambda}\right)^j}{2^{i/2}} \dfrac{j!}{(j/2)!}, & \text{if } j \text{ is even}, \\ 0, & \text{if } j \text{ is odd}. \end{cases}$$

(23)

Here

$$V\left(z_\eta^{i,\Lambda} \middle| H_i\right) = \vartheta_{z,\eta}^{i,\Lambda}$$

$$= \sum_{\ell_1=1}^{\Lambda} \sum_{\ell_2=1}^{\Lambda} C_{\eta,\ell_1}^{i,\Lambda} D_{\ell_2}^{i,p_{\ell_2}} \times \left(\rho_{\ell_1,\ell_2}^i + b_{\ell_1}^{i,p_{\ell_1}} b_{\ell_2}^{i,p_{\ell_2}}\right)$$

$$-\left(\sum_{\ell=1}^{\Lambda} C_{\eta,\ell}^{i,\Lambda} b_{\ell}^{i,p_{\ell}} - D_{\ell}^{i,p_{\ell}}\right)^2,$$

$$C_{\eta,\ell}^{i,\Lambda} = \sum_{\delta=1}^{\Lambda} \beta_{\eta,\delta}^i K_{\delta,\ell}^i,$$

$$D_{\ell}^{i,p_{\ell}} = \left(\sum_{\delta=1}^{\Lambda} \beta_{\eta,\delta}^i K_{\delta,\ell}^i\right) b_{\ell}^{i,p_{\ell}}.$$

(24)

Taking advantage of ratios (21), (22), for computation of the moments (13), we obtain expression (14).

Probability integral (10) can be computed with the help of Edgewort's series [19-21] using formula (14) for computation of the initial moments of random variable (9). In particular, in the considered case, using wellknown techniques of obtaining Edgewort's series [30], we have

$$\int_{r}^{0} p_j\left(z_1 \middle| H_i\right) dz \approx \hat{\pi}_j\left(0 \middle| H_i\right) = \Phi\left(z_1^*\right)$$

$$+ \left\{\frac{1}{3!} \frac{[\varsigma_3]_j}{\left(\mu_2^{j,i}\right)^{3/2}} \times \left(z_1^{*3} - 3z_1^*\right) + \frac{1}{4!} \frac{[\varsigma_4]_j}{\left(\mu_2^{j,i}\right)^2} \left(z_1^{*4} - 6z_1^{*2} + 3\right)\right.$$

$$+\frac{10}{6!}\left(\frac{\left[\varsigma_3\right]_j^i}{\left(\mu_2^{j,i}\right)^{3/2}}\right)^2\left(z_1^{*6}-15z_1^{*4}+45z_1^{*2}-15\right)$$

$$+\frac{1}{5!}\frac{\left[\varsigma_5\right]_j^i}{\left(\mu_2^{j,i}\right)^{5/2}}\left(z_1^{*5}-10z_1^{*3}+15z_1^{*}\right)-\frac{35}{7!}\frac{\left[\varsigma_3\right]_j^i\left[\varsigma_4\right]_j^i}{\left(\mu_2^{j,i}\right)^{7/2}}$$

$$\times\left(z_1^{*7}-21z_1^{*5}+105z_1^{*3}-105z_1^{*}\right)-\frac{280}{9!}\left(\frac{\left[\varsigma_3\right]_j^i}{\left(\mu_2^{j,i}\right)^{3/2}}\right)^3$$

$$\times\left(z_1^{*9}-36z_1^{*7}+378z_1^{*5}-1260z_1^{*3}+945z_1^{*}\right)$$

$$+\frac{1}{6!}\frac{\left[\varsigma_6\right]_j^i}{\left(\mu_2^{j,i}\right)^3}\left(z_1^{*6}-15z_1^{*4}+45z_1^{*2}-15\right)+\frac{35}{8!}\left(\frac{\left[\varsigma_4\right]_j^i}{\left(\mu_2^{j,i}\right)^2}\right)^2$$

$$\times\left(z_1^{*8}-28z_1^{*6}+210z_1^{*4}-420z_1^{*2}+105\right)$$

$$+\frac{2100}{10!}\frac{\left[\varsigma_3^2\right]_j^i\left[\varsigma_4\right]_j^i}{\left(\mu_2^{j,i}\right)^5}\times\left(z_1^{*10}-45z_1^{*8}+630z_1^{*6}\right.$$

$$-3150z_1^{*4}+4725z_1^{*2}-945\bigg)+\frac{23100}{12!}\left(\frac{\left[\varsigma_3\right]_j^i}{\left(\mu_2^{j,i}\right)^{3/2}}\right)^4$$

$$\times\left(z_1^{*12}-66z_1^{*10}+1485z_1^{*8}-13860z_1^{*6}\right.$$

$$+51975z_1^{*4}-62370z_1^{*2}+10395\big)\big\}\,\alpha\left(z_1^{*}\right),\qquad(25)$$

Where $z_k^* = \left(-\mu_1^{j,i}\right)/\sqrt{\mu_2^{j,i}}; [\varsigma_k]_j^i, (k = 1, 2, ...)$, is the k th semi-invariant of the random variable ξ_j provided hypothesis H_i is true (the computation of semi-invariants is not difficult knowing all initial moments include-ing $\mu_k^{j,i}$ (see, for example, [21])); $\dot{\mu}_2^{j,i}$ is the second central moment; $\alpha(x)$ is standard normal density, i.e.

$$\alpha(x) = \exp\left(-x^2/2\right)/\sqrt{2\pi} .$$

Satisfying the first seven terms in expansion (25), the absolute value of calculation error of the probability integral is calculated by the formula

$$\left|\Delta_i^j\right| \leq \left\|\left[\frac{1}{6!}\frac{[\varsigma_6]_j^i}{\left(\mu_2^{j,i}\right)^3}\left(z_1^{*6} - 15z_1^{*4} + 45z_1^{*2} - 15\right)\right.\right.$$

$$+\frac{35}{8!}\left(\frac{[\varsigma_4]_j^i}{\left(\mu_2^{j,i}\right)^2}\right)^2\left(z_1^{*8} - 28z_1^{*6} + 210z_1^{*4} - 420z_1^{*2} + 105\right)$$

$$+\frac{2100}{10!}\frac{\left[\varsigma_3^2\right]_j^i[\varsigma_4]_j^i}{\left(\mu_2^{j,i}\right)^5}\left(z_1^{*10} - 45z_1^{*8} + 630z_1^{*6} - 3150z_1^{*4}\right.$$

$$+4725z_1^{*2} - 945\right) + \frac{23100}{12!}\left(\frac{[\varsigma_3]_j^i}{\left(\mu_2^{j,i}\right)^{3/2}}\right)^4$$

$$\left(z_1^{*12} - 66z_1^{*10} + 1485z_1^{*8} - 13860z_1^{*6} + 51975z_1^{*4}\right.$$

$$\left.\left.\left.-62370z_1^{*2} + 10395\right\}\right)\alpha\left(z_1^*\right)\right| .$$

The variable $\xi_j = \sum_{\ell=1}^{s} C_\ell^j \exp(-y_\ell)$ is continuous and unambiguously defined for every value of x. Therefore, the random variable ξ_j is continuous. The characteristic function of the random variable ξ_j and its derivatives of any order exist, as the moments of any order of this random variable exist. At the same time, the characteristic function is uniformly continuous. Consequently, the distribution function of this random variable exists and is continuous [21].

Theorem 3.2:

The distribution function of random variable ξ_j exists and is uniquely determined by moments (14).

Proof:

For proving this theorem, it is necessary to show that all moments $\mu_r^{j,i}, r = 1, 2, \ldots$ exist and the following condition takes place [19,21]

$$\limsup_{r \to \infty} \frac{\left(\mu_{2r}^{j,i} \right)^{1/2r}}{2r} < \infty .$$

The fact that all moments exist is obvious from formula (14) as by using it, it is possible to calculate the moments of any order with any specified accuracy. The values of these moments exist and are finite.

When solving the practical problems coefficients k_ℓ^k take on the values bounded above; correlation matrices W_ℓ are positively determined matrices the determinants of which differ from zero. Therefore, coefficients $\left| C_\ell^j \right|$ are bounded-above quantities.

There takes place

$$E\left[\exp\left(-\left(y_{\ell_1} + \cdots + y_{\ell_r} \right) \right) \middle| H_i \right] = \int_{-\infty}^{+\infty} \cdots \int_{-\infty}^{+\infty} e^{-\left(y_{\ell_1} + \cdots + y_{\ell_r} \right)}$$

$$\times N\left(\mathbf{x} \middle| \mathbf{a}^i, \mathbf{W}_i \right) d\mathbf{x},$$

where $y_{\ell_1} + \ldots + y_{\ell_r}$ is the sum of quadratic forms of normally distributed n-dimensional vector x at different vectors of mathematical expectations and covariance matrices. Therefore, at changing components of the vector x from $-\infty$ up to $+\infty$, the quadratic form $y_{\ell_1} + \ldots + y_{\ell_r}$ takes the values from 0 to $+\infty$, and the value of function $e^{-(y_{\ell_1} + \ldots + y_{\ell_r})}$ varies from 1 to 0, respectively. Therefore [18]

$$E\left[\exp\left(-\left(y_{\ell_1} + \cdots + y_{\ell_r}\right)\right) \middle| H_i \right]$$

$$\leq \int_{-\infty}^{+\infty} \cdots \int_{-\infty}^{+\infty} N\left(x \middle| a^i, W_i\right) dx = 1$$

Thus, taking into account (5) and (11) we can write down

$$\left| \mu_r^{j,i} \right| \leq \left| \sum_{\ell_1=1}^{S} \cdots \sum_{\ell_r=1}^{S} C_{\ell_1}^j \cdots C_{\ell_r}^j \right|,$$

$$\leq A^r \left(2\pi\right)^{-\frac{nr}{2}} \sum_{\ell_1=1}^{S} \cdots \sum_{\ell_r=1}^{S} \prod_{v=1}^{r} p\left(H_{\ell_v}\right) \left| W_{\ell_v} \right|^{-1}$$

where A is the maximum by absolute value among coefficients k_ℓ^j.

Assume r=1, then we have

$$\left| \mu_1^{j,i} \right| \leq A (2\pi)^{-\frac{n}{2}} \sum_{\ell=1}^{S} p\left(H_\ell\right) \left| W_\ell \right|^{-1}.$$

Let us designate $\left| W_{min} \right| = \min_{\{\ell\}} \left| W_\ell \right|$. Then

$$\mu_1^{j,i} \leq A (2\pi)^{-\frac{n}{2}} \left| W_{min} \right|^{-1} \sum_{\ell=1}^{S} p\left(H_\ell\right) = A (2\pi)^{-\frac{n}{2}} \left| W_{min} \right|^{-1}.$$

If $A \leq (2\pi)^{\frac{n}{2}} |W_{min}|$, then $\mu_1^{j,i} \leq 1$ and it is not difficult to be convinced that $\mu_r^{j,i} \to 0$ at $r \to \infty$.

Let $A = C(2\pi)^{\frac{n}{2}} |W_{min}|$, where C>1. Hence

$\mu_r^{j,i} \leq C^r Sr$ and

$$\frac{\left(\mu_{2r}^{j,i}\right)^{1/2r}}{2r} \leq \frac{\left(C^{2r}S2r\right)^{1/2r}}{2r} = \frac{CS^{1/2r}(2r)^{1/2r}}{2r}$$

$$= CS^{1/2r}(2r)^{-\frac{2r-1}{2r}} = \frac{CS^{1/2r}}{(2r)^{\frac{2r-1}{2r}}} \to \frac{C}{2r} \to 0$$

at $r \to \infty$, which proves the theorem.

COMPUTATION RESULTS

The accuracy of this algorithm depends heavily on M- the number of used items in expansion (12). In order to increase the accuracy of approximation of the exponent for given M and, in general, the reliability of computation in the tasks of hypotheses testing, it is expedient to perform first the normalization of initial data by formulae:

$$x_i' = (x - c_i)/(d_i - c_i),$$

$$\rho_{ij}' = \rho_{ij}/\left[(d_i - c_i)\cdot(d_j - c_j)\right], i, j = 1, \cdots, n$$

Where c_i, d_i, i=1,...,n, are the minimum and maximum values of the i th parameter for the given set of the considered hypotheses, i.e.

$$c_i = \min_{\{j\}}\{a_i^j\}, \quad d_i = \max_{\{j\}}\{a_i^j\},$$

i=1,...,n, j=1,...,S [3]. In this case, the values of the parameters of the algorithm M=15 and seven items in expansion (25) provided the absolute error of computation of integral (1) that does not exceed 0.005 for computed examples (see below). This fact was established by modeling for the observation vector with noncorrelated components. Unfortunately, by now the considered algorithm has been realized only for such a case [31].

The results of simulation showed that the time of execution of the task (decision-making and computation of the suitable value of the risk function) by using the Monte-Carlo method made up 1.2 sec, sec, and sec for the number of hypotheses S=3, S=4 and S=5, respectively, the dimensionality of the observed vector being equal to n=8 in all cases. The tested hypotheses and correlation matrix for the case S=5 are given in tables of Figure 1 and Figure 2, respectively. Figures are presented as the suitable forms of the task of hypotheses test of the statistical software in which the appropriate methods are realized [31]. For other values of S, there are chosen the suitable sub-sets of the tables of these Figures. In the first column of the table of Figure 1 is given the vector of measurements and in the other columns are given hypothetical values of mathematical expectation of this vector.

Meanwhile, when using the method offered here, the computation time did not practically change and the results were obtained for the time noticeably less than 1 sec. In both cases probability integrals (7) were computed with the accuracy of ≤0.005. In Figures 3 and 4 are given the dependences of the integral computation time on the accuracy and number of tested hypotheses respecttively.

At solving many practical problems, especially military problems [5,32], the dimensionality of the integrals like (1) often is equal to several tens and difference between the computation time necessary for the considered methods is significantly longer than in the above mentioned case [14], whereas the computation time for solving the defence problems are of great importance.

j	x_j	a_{1j}	a_{2j}	a_{3j}	a_{4j}	a_{5j}
1	2.010000	1.010000	4.010000	7.010000	10.01000	13.00000
2	2.020000	1.020000	4.020000	7.020000	10.02000	13.00000
3	2.030000	1.030000	4.030000	7.030000	10.03000	13.00000
4	2.040000	1.040000	4.040000	7.040000	10.04000	13.00000
5	2.050000	1.050000	4.050000	7.050000	10.05000	13.00000
6	2.060000	1.060000	4.060000	7.060000	10.06000	13.00000
7	2.070000	1.070000	4.070000	7.070000	10.07000	13.00000
8	2.080000	1.080000	4.080000	7.080000	10.08000	13.00000

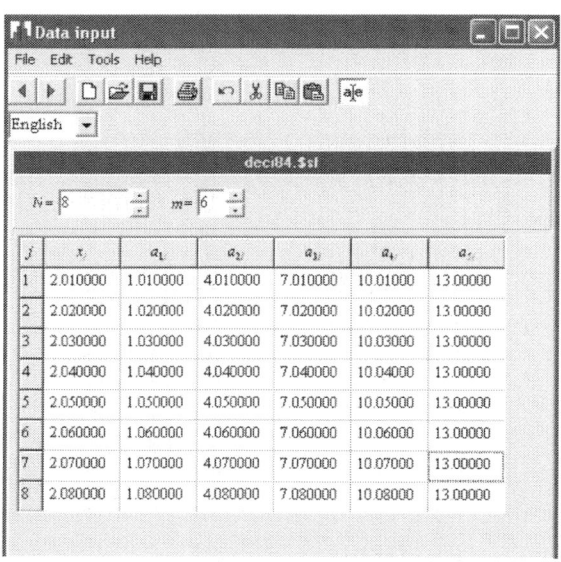

Figure 1: The form of entering the tested hypotheses and a measurement vector.

j\i	1	2	3	4	5	6	7	8
1	1.79560	1.36360	1.93160	2.49960	3.06760	3.63560	4.20360	4.77160
2	1.36360	2.40320	1.97920	2.55520	3.13120	3.70720	4.28320	4.85920
3	1.93160	1.97920	3.02680	2.61080	3.19480	3.77880	4.36280	4.94680
4	2.49960	2.55520	2.61080	3.66640	3.25840	3.85040	4.44240	5.03440
5	3.06760	3.13120	3.19480	3.25840	4.32200	3.92200	4.52200	5.12200
6	3.63560	3.70720	3.77880	3.85040	3.92200	4.99360	4.60160	5.20960
7	4.20360	4.28320	4.36280	4.44240	4.52200	4.60160	5.68120	5.29720
8	4.77160	4.85920	4.94680	5.03440	5.12200	5.20960	5.29720	6.38480

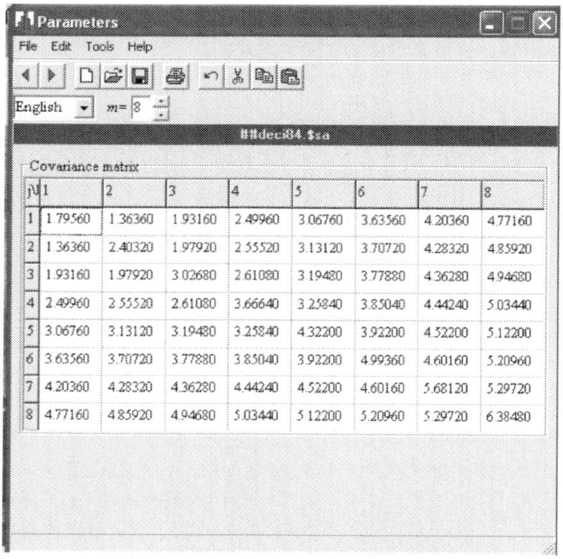

Figure 2: The form of entering the covariance matrix.

The theoretical investigation of the dependence of the accuracy of computation of integral (1) on M-the number of items in expansion (13) is a challenging task. Therefore, at program realization of the offered algorithm and, in general, algorithms of such a kind, it is worthwhile to make parameter M and the number of items in expansion (25) external parameters of the program. This allows establishing their optimal values for each concrete case by experimentation depending on the desired accuracy of computation.

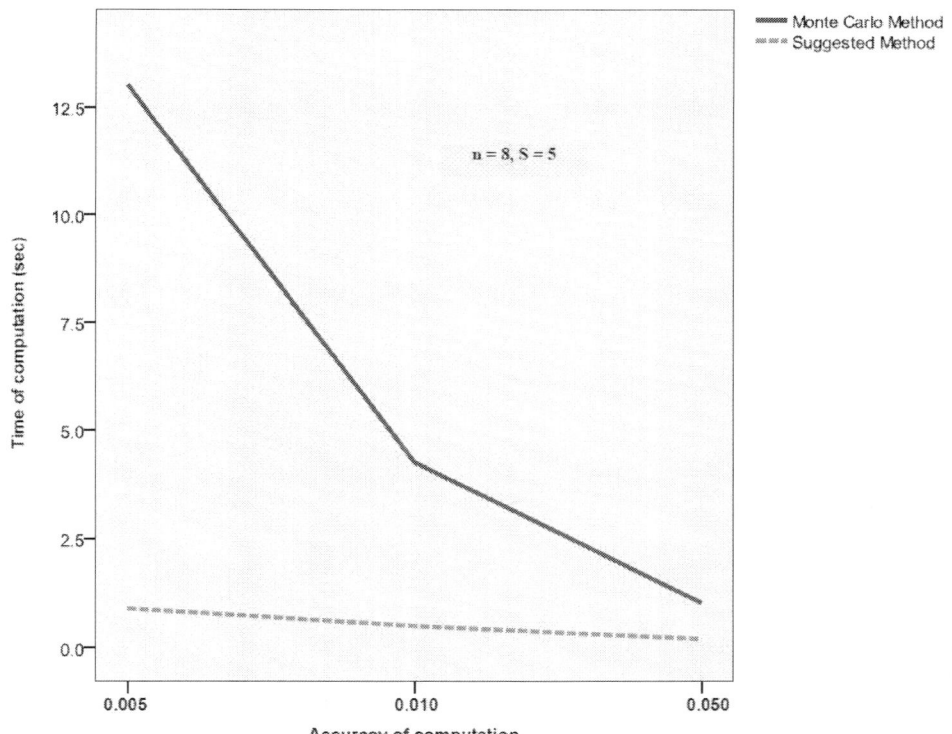

Figure 3: Dependence of the integral computation time on the accuracy.

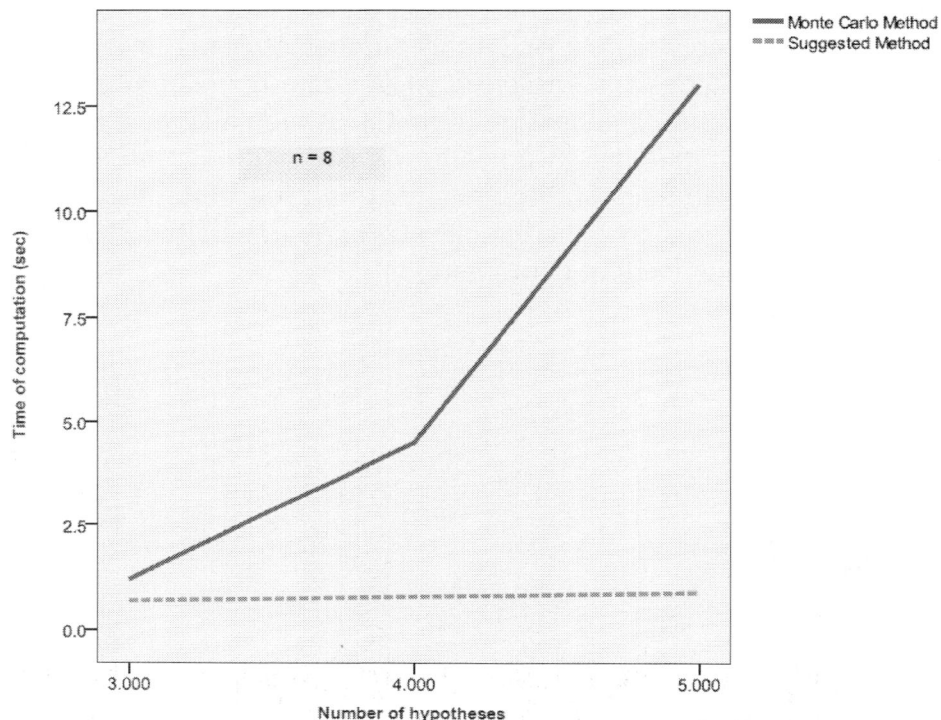

Figure 4: Dependence of the integral computation time on the number of hypotheses.

CONCLUSIONS

The method of computation of the probability integral from the multivariate normal density over the Complex Subspace by using series and the reduction of dimensionality of the multidimensional integral to one without losing the information was developed. The formulae for computation of product moments of normalized normally distributed random variables were also derived. The existence of the probability distribution law of the weighted sum of exponents of negative quadratic forms of the normally distributed random vector was justified. The opportunity of its unambiguous determination by the computed moments was proved.

REFERENCES

1. S. Thompson, "On the Distribution of Type II Errors in Hypothesis Testing," Applied Mathematics, Vol. 2, No. 2, 2011, pp. 189-195. doi:10.4236/am.2011.22021
2. C. R. Rao, "Linear Statistical Inference and Its Application," 2nd Edition, John Wiley & Sons Ltd, New York, 2006.
3. K. J. Kachiashvili, "Generalization of Bayesian Rule of Many Simple Hypotheses Testing," International Journal of Information Technology & Decision Making, Vol. 2, No. 1, 2003, pp. 41-70. doi:10.1142/S0219622003000525
4. A. V. Primak, V. V. Kafarov and K. J. Kachiashvili, "System Analysis of Air and Water Quality Control," Naukova Dumka, Kiev, 1991.
5. A. I. Potapov, A. G. Vinogradov, I. A. Goritskyi and E. E. Pertsov, "About Decision-Making of Presence of Objects at Group Measurements," Questions of Radio-Electronics, Vol. 6, 1975, pp. 69-76.
6. P. J. David and P. Rabinovitz, "Methods of Numerical Integration. Computer Science and Applied Mathematics," 2nd Edition, Academic Press Inc., Orlando, 1984.
7. A. Genz, "Numerical Computation of Multivariate Normal Probabilities," Journal of Computational and Graphical Statistics, Vol. 1, 1992, pp. 141-149.
8. A. Genz, "Comparison of Methods for the Computation of Multivariate Normal Probabilities," Computing Science and Statistics, Vol. 25, 1993, pp. 400-405.
9. A. Genz and F. Bretz, "Numerical Computation of Multivariate t-Probabilities with Application to Power Calculation of Multiple Contrasts," Journal of Statistical Computation and Simulation, Vol. 63, No. 4, 1999, pp. 361- 378. doi:10.1080/00949659908811962
10. S. Joe, "Approximations to Multivariate Normal Rectangle Probabilities Based on Conditional Expectations," Journal of the American Statistical Association, Vol. 90, 1995, pp. 957-964.
11. I. H. Sloan and S. Joe, "Lattice Methods for Multiple Integration," Clarendon Press, Oxford, 1994.
12. V. Hajivassiliou, D. McFadden and P. Ruud, "Simulation of Multivariate Normal Rectangle Probabilities and Their Derivatives: Theoretical and Computational Results," Journal of Econometrics, Vol. 72, No. 1-2, 1996, pp. 85- 134. doi:10.1016/0304-4076(94)01716-6
13. J. O. Berger, "Statistical Decision Theory and Bayesian Analysis," Springer, New York, 1985.
14. K. J. Kachiashvili, "Bayesian Algorithms of Many Hypothesis Testing," Ganatleba, Tbilisi, 1989.
15. D. V. Lindley, "The Use of Prior Probability Distributions in Statistical Inference and Decisions," Proceedings of the 4th Berkeley Symposium on Mathematical Statistics and Probability, Vol. 1, 1961, pp. 453-468.
16. L. Tierney and J. B. Kadane, "Accurate Approximations for Posterior Moments and Marginal Densities," Journal of the American Statistical Association, Vol. 81, 1986, pp. 82-86.

17. A. Stuart, J. K. Ord and S. Arnols, "Kendall's Advanced Theory of Statistics. Classical Inference and the Linear Model," 6th Edition, Vol. 2A, Oxford University Press Inc., New York, 1999.

18. T. W. Anderson, "An introduction to Multivariate Statistical Analysis," 3rd Edition, Wiley & Sons, Inc., New Jersey, 2003.

19. A. Stuart, J. K. Ord and S. Arnols, "Kendall's Advanced Theory of Statistics. Distribution Theory," 6th Edition, Vol. 1, Oxford University Press Inc., New York, 1994.

20. H. Cramer, "Mathematical Methods of Statistics," Princeton University Press, Princeton, 1999.

21. M. Kendall and A. Stuart, "Distribution Theory," Vol. 1, Charles Griffit & Company Limited, London, 1966.

22. G. D. Shellard, "Estimating the Product of Several Random Variables," Journal of the American Statistical Association, Vol. 47, 1952, pp. 216-221.

23. H. A. R. Barnett, "The Variance of the Product of Two independent Variables and Its Application to an Investigation Based on Sample Data," Journal of the Institute of Actuaries, Vol. 81, 1955, pp. 190-198.

24. L. A. Goodman, "On the Exact Variance of Products," Journal of the American Statistical Association, Vol. 55, 1960, pp. 708-713.

25. L. A. Goodman, "The Variance of the Product of K Random Variables," Journal of the American Statistical Association, Vol. 57, No. 297, 1962, pp. 54-60.

26. S. N. Nath, "On Product Moments from a Finite Universe," Journal of the American Statistical Association, Vol. 63, No. 322, 1968, pp. 535-541.

27. S. N. Nath, "More Results on Product Moments from a Finite Universe," Journal of the American Statistical Association, Vol. 64, No. 327, 1969, pp. 864-869.

28. S. Nadarajah and K. Mitov, "Product Moments of Multivariate Random Vectors," Communications in Statistics. Theory and Methods, Vol. 32, No. 1, 2003, pp. 47-60.doi:10.1081/STA-120017799

29. S. Kotz, N. Balakrishnan and N. L. Johnson, "Continuous Multivariate Distributions. Models and Applications," Vol. 1, 2nd Edition, John Wiley & Sons Ltd, New York, 2000.doi:10.1002/0471722065

30. G. Szego, "Orthogonal Polynomials," American Mathematical Society, New York, 1959.

31. K. J. Kachiashvili and D. I. Melikdzhanian, "SDpro—The Software Package for Statistical Processing of Experimental Information," International Journal Information Technology & Decision Making, Vol. 9, No 1, 2010, pp. 1-30. doi:10.1142/S0219622010003634

32. K. J. Kachiashvili and A. Mueed "Conditional Bayesian Task of Testing Many Hypotheses," Statistics, 2011, pp. 1-20. doi:10.1080/02331888.2011.602681

CITATION

K. Kachiashvili and M. Hashmi, "Computation of the Multivariate Normal Integral over a Complex Subspace,"Applied Mathematics, Vol. 3 No. 5, 2012, pp. 489-498. doi: 10.4236/am.2012.35074.

Chebyshev Polynomials for Solving a Class of Singular Integral Equations

Samah M. Dardery[1] and Mohamed M. Allan[2]

[1]Department of Mathematics, Faculty of Science, Zagazig University, Zagazig, Egypt
[2]Department of Mathematics, Faculty of Science and Arts Al-Mithnab, Qassim University, Qassim, KSA

ABSTRACT

This paper is devoted to study the approximate solution of singular integral equations by means of Chebyshev polynomials. Some examples are presented to illustrate the method.

INTRODUCTION

During the last three decades, the singular integral equation methods with applications to several basic fields of engineering mechanics, like elasticity, plasticity, aerodynamics and fracture mechanics have been studied and improved by several scientists (see [1] -[6]). Hence, it is interest to solve numerically this type of integral equations (see [7] [8]). Chebyshev polynomials are of great importance in many areas of mathematics particularly approximation theory (see [9] [10]).

In this paper we analyze the numerical solution of singular integral equations by using Chebyshev polynomials of first, second, third and fourth kind to obtain systems of linear algebraic equations, these systems are solved numerically. The methodology of the present work expected to be useful for solving singular integral equations of the first kind, involving partly singular and partly regular kernels. The singularity is assumed to be of the Cauchy type. The method is illustrated by considering some examples.

Singular integral equation of first kind, with a Cauchy type singular kernel, over a finite interval can be represented by

$$\int_{-1}^{1} \frac{k(t,x)\varphi(t)}{t-x}dt + \int_{-1}^{1} L(t,x)\varphi(t)dt = f(x), -1 < x < 1$$

(1.1)

where k(t, x), L(t, x) and f(x) are given real-valued continuous functions belonging to the class Holder of continues functions and $k(t,t) \neq 0$. In Equation (1.1) the singular kernel is interpreted as Cauchy principle value. Integral equation of form (1.1) and other different forms have many applications (see [1] [2] [6] [11] [12]). The theory of this equation is well known and it is presented in [13] [14]. An approximate method for solving (1.1) using a polynomial approximation of degree n has been proposed in [7].

It is well known that the analytical solutions of the simple singular integral equation

$$\int_{-1}^{1} \frac{\varphi(t)}{t-x}dt = f(x), -1 < x < 1$$

(1.2)

at k(t, x)=1 and L(t, x)=0, for the following four cases:

1) The solution is unbounded at both end-points $x = \pm 1$,

2) The solution is bounded at both end-points $x = \pm 1$,

3) The solution is bounded at end $x = -1$, but unbounded at end $x = +1$, 4) The solution is unbounded at end $x = -1$, but bounded at end $x = +1$, are given by [15] . In this paper the used approximate method for solving Equation (1.1) stems from recent work [10] wherein an approximate method has been developed to solve

the simple Equation (1.2). The approximate method developed below appears to be quite appropriate for solving the most general type Equation (1.1). Some examples are presented to illustrate the method.

THE APPROXIMATE SOLUTION

In this section we present the method of the approximate solution of Equation (1.1) in four cases.

Let the unknown function φ in Equation (1.1) be approximated by the polynomial function

$$\varphi_n(x) = W^{(j)}(x)\sum_{i=0}^{n} c_i^{(j)}\Psi_i^{(j)}(x), (j = 1, 2, 3, 4) \tag{2.1}$$

where $c_i^{(i)}, i = 0, 1, 2, \cdots, n$ are unknown coefficients, to be determined, and in case (I): $\psi_i^{(1)}(x) = T_i(x)$, in case (II): $\psi_i^{(2)}(x) = u_i(x)$, in case (III): $\psi_i^{(3)}(x) = v_i(x)$ and in case (VI): $\psi_i^{(4)}(x) = w_i(x)$, where t_i, u_i, v_i, ,and $w_i, i = 0, 1, \ldots.n$, are The Chebyshev polynomials of the first, second, third and fourth kinds respectively can be defined by the recurrence relations [9] [16] .

$$\left.\begin{aligned} &T_0(x) = 1, T_1(x) = x \\ &T_n(x) = 2xT_{n-1}(x) - T_{n-2}(x), n \geq 2 \end{aligned}\right\} \tag{2.2}$$

$$\left.\begin{aligned} &U_0(x) = 1, U_1(x) = 2x \\ &U_n(x) = 2xU_{n-1}(x) - U_{n-2}(x), n \geq 2 \end{aligned}\right\} \tag{2.3}$$

$$\left.\begin{aligned} &V_0(x) = 1, V_1(x) = 2x - 1 \\ &V_n(x) = 2xV_{n-1}(x) - V_{n-2}(x), n \geq 2 \end{aligned}\right\} \tag{2.4}$$

$$W_0(x) = 1, W_1(x) = 2x + 1$$
$$W_n(x) = 2xW_{n-1}(x) - W_{n-2}(x), n \geq 2$$

$$(2.5)$$

and $w^i, i = 0, 1, \ldots, n$, are the corresponding weight functions.

Substituting the approximate solution (2.1) for the unknown function into (1.1) yields

$$\sum_{i=0}^{n} c_i^{(j)} \left[\int_{-1}^{1} \frac{k(t,x)W^{(j)}(t)\Psi_i^{(j)}(t)}{t-x} dt + \int_{-1}^{1} L(t,x)W^{(j)}(t)\Psi_i^{(j)}(t) dt \right] = f(x), -1 < x < 1$$

$$(2.6)$$

In above Equation (2.6), we next use the following Chebyshev approximation to the kernels k (t, x) and L(t, x), given by (for fixed x, cf. [7])

$$k(t,x) \cong \sum_{p=0}^{m} k_p(x)t^p, L(t,x) \cong \sum_{q=0}^{s} L_q(x)t^q$$

$$(2.7)$$

with known expressions for $K_p(x)$ and $L_q(x)$. Then (2.6) gives

$$\sum_{i=0}^{n} c_i^{(j)}\alpha_i^{(j)}(x) = f(x), -1 < x < 1, (j = 1, 2, 3, 4)$$

$$(2.8)$$

where

$$\alpha_i^{(j)}(x) = \sum_{p=0}^{m} k_p(x)u_{p,i}^{(j)}(x) + \sum_{q=0}^{s} L_q(x)\gamma_{q,i}^{(j)}$$

$$(2.9)$$

with

$$u_{p,i}^{(j)}(x) = \int_{-1}^{1} \frac{t^p W^{(j)}(t)\Psi_i^{(j)}(t)}{t-x} dt, -1 < x < 1, (j = 1, 2, 3, 4)$$

$$(2.10)$$

and

$$\gamma_{q,i}^{(j)} = \int_{-1}^{1} t^q W^{(j)}(t) \Psi_i^{(j)}(t) \, dt \tag{2.11}$$

Let $x_k^{(j)}, j = 1, 2, 3, 4$, be the zeros of $U_n(x), T_{n+2}(x), W_{n+1}(x)$ and $V_{n+1}(x)$, respectively. Substituting the collocation points $x_k^{(j)}, j = 1, 2, 3, 4$ into (2.8) we obtain the following systems of linear equations:

$$\sum_{i=0}^{n} c_i^{(j)} \alpha_i^{(j)} \left(x_k^{(j)} \right) = f \left(x_k^{(j)} \right), (k = 1, 2, \cdots, n+1), (j = 1, 2, 3, 4) \tag{2.12}$$

where

$$\alpha_i^{(j)} \left(x_k^{(j)} \right) = \sum_{p=0}^{m} k_p \left(x_k^{(j)} \right) u_{p,i}^{(j)} \left(x_k^{(j)} \right) + \sum_{q=0}^{s} L_q \left(x_k^{(j)} \right) \gamma_{q,i}^{(j)} . (k = 1, 2, \cdots, n+1), (j = 1, 2, 3, 4) \tag{2.13}$$

Solving the system of Equation (2.12) for the unknown coefficients $C_i^{(j)}, j = 1, 2, 3, 4$, and substituting the values of $C_i^{(j)}$ into (2.1) we obtain the approximate solutions of Equation (1.1).

NUMERICAL EXAMPLES

In this section, we consider some problems to illustrate the above method. All results were computed using FORTRAN code.

Example 1:

Consider the following singular integral equation

$$\int_{-1}^{1} \frac{\left(x + t^2 \right) \varphi(t)}{t - x} + \int_{-1}^{1} \left(x^2 + t^3 \right) \varphi(t) \, dt = 2x^4 - 2x^2 - \frac{3}{8}, -1 < x < 1 \tag{3.1.1}$$

where $k(x,t) = x + t^2.L(x,t) = x^2 + t^3, f(t) = 2x^4 - 2x^2 - \dfrac{3}{8}$. So, one gets

$$k_0(x) = x, k_1(x) = 0, k_2(x) = 1, k_p(x) = 0, (p > 2)$$

$$L_0(x) = x^2, L_1(x) = 0, L_2(x) = 0, L_3(x) = 1, L_q(x) = 0, (q > 3)$$

Hence we find that relation (2.8) produces

$$\sum_{i=0}^{n} c_i^{(j)} \alpha_i^{(j)}(x) = 2x^4 - 2x^2 - \frac{3}{8}, -1 < x < 1, (j = 1,2,3,4)$$

Thus (2.9) gives

$$\alpha_i^{(j)}(x) = xu_{0,i}^{(j)}(x) + u_{2,i}^{(j)}(x) + x^2\gamma_{0,i}^{(j)} + \gamma_{3,i}^{(j)}, (j = 1,2,3,4), (i = 0,1,2,\cdots)$$

Firstly, let us consider in detail the case (I), j=1, for n=3. This results in

$$u_{0,i}^{(1)}(x) = \int_{-1}^{1} \frac{T_i(t)}{\sqrt{1-t^2}\, t - x}\, dt, u_{2,i}^{(1)}(x) = \int_{-1}^{1} \frac{t^2 T_i(t)}{\sqrt{1-t^2}\, t - x}\, dt, -1 < t < 1,$$

$$\tag{3.1.2}$$

$$\gamma_{0,i}^{(1)} = \int_{-1}^{1} \frac{T_i(t)}{\sqrt{1-t^2}}\, dt, \quad \gamma_{3,i}^{(1)} = \int_{-1}^{1} \frac{t^3 T_i(t)}{\sqrt{1-t^2}}\, dt,$$

$$\tag{3.1.3}$$

By applying the following relations

$$\int_{-1}^{1} \frac{T_i(t)}{\sqrt{1-t^2}\,(t-x)}\, dt = \pi U_{i-1}(x), \int_{-1}^{1} \frac{1}{\sqrt{1-t^2}\,(t-x)}\, dt = 0$$

$$\tag{3.1.4}$$

$$\int_{-1}^{1} \frac{T_i(t)T_j(t)}{\sqrt{1-t^2}} dt = \begin{cases} 0 & i \neq j \\ \pi & i = j = 0 \\ \pi/2 & i = j \neq 0 \end{cases}$$

$$(3.1.5)$$

It is easy to estimate the values $u_{0,i}^{(1)}, u_{2,i}^{(1)}, \gamma_{0,i}^{(1)}$ and $\gamma_{3,i}^{(1)}$.

From (2.2) and (3.1.2)-(3.1.5) we get

$$\alpha_i^{(1)}(x) = \begin{cases} \pi(x^2 + x); & i = 0 \\ \pi\left(x^2 + x + \dfrac{7}{8}\right); & i = 1 \\ \pi(2x^3 + 2x^2); & i = 2 \\ \pi\left(4x^4 + 4x^3 - x^2 - x + \dfrac{1}{8}\right); & i = 3 \end{cases}$$

$$(3.1.6)$$

By choosing the collocation points

$$x_k = \cos\left(\frac{(2k-1)\pi}{2(n+2)}\right), (k = 1, 2, 3, 4)$$

for n=3 we obtain the following system of linear equations:

$$\sum_{i=0}^{3} c_i^{(1)} \alpha_i^{(1)}(x_k) = f(x_k), k = 1, 2, 3, 4$$

By solving this system for the unknown coefficients $C_i^{(1)}, i = 0, 1, 2, 3$ that produces

$$c_0^{(1)} = 0.3183098, c_1^{(1)} = -0.1591549$$
$$c_2^{(1)} = -0.3183098, c_3^{(1)} = 0.1591549$$

$$(3.1.7)$$

From (3.1.7) we obtain the approximate solution of Equation (3.1.1) in the form

$$\varphi_n(x) \cong \frac{2}{\pi\sqrt{1-x^2}}\left(x^3 - x^2 - x + 1\right)$$

$$(3.1.8)$$

Which coincides with the exact solution. The error of approximate solution (3.1.8) of Equation (3.1.1) at n=20 is given by Table 1.

Secondly, let us consider in detail the case (II), j=2, for n=3. This results in

$$u_{0,i}^{(2)}(x) = \int_{-1}^{1} \frac{\sqrt{1-t^2}\,U_i(t)}{t-x}\,dt, u_{2,i}^{(2)}(x) = \int_{-1}^{1} \frac{t^2\sqrt{1-t^2}\,U_i(t)}{t-x}\,dt, -1 < t < 1,$$

$$(3.1.9)$$

$$\gamma_{0,i}^{(2)} = \int_{-1}^{1} \sqrt{1-t^2}\,U_i(t)\,dt, \quad \gamma_{3,i}^{(2)} = \int_{-1}^{1} \sqrt{1-t^2}\,t^3 U_i(t)\,dt,$$

$$(3.1.10)$$

By applying the following relations

$$\int_{-1}^{1} \frac{\sqrt{1-t^2}\,U_i(t)}{t-x}\,dt = -\pi T_{i+1}(x)$$

$$(3.1.11)$$

$$\int_{-1}^{1} \sqrt{1-t^2}\,U_i(t)U_j(t)\,dt = \begin{cases} 0 & i \neq j \\ \dfrac{\pi}{2} & i = j \end{cases}$$

$$(3.1.12)$$

It is easy to estimate the values $u_{0,i}^{(1)}, u_{2,i}^{(1)}, \gamma_{0,i}^{(1)}$ and $\gamma_{3,i}^{(1)}$.

From the relations (2.3) and (3.1.9)-(3.1.12) we get

$$\alpha_i^{(2)}(x) = \begin{cases} -\dfrac{\pi}{2}\left(2x^3 + x^2 - x\right) \\[2mm] -\pi\left(2x^4 + 2x^3 - x^2 - x - \dfrac{3}{8}\right) \\[2mm] -\pi\left(4x^5 + 4x^4 - 3x^3 - 3x^2 + \dfrac{1}{4}x\right) \\[2mm] -\pi\left(8x^6 + 8x^5 - 8x^4 - 8x^3 + x^2 + x - \dfrac{1}{16}\right) \end{cases}$$

(3.1.13)

By choosing the collocation points

$$x_k^{(2)} = \cos\left(\frac{(2k-1)\pi}{2(n+2)}\right), (k = 1, 2, 3, 4)$$

for n=3 we obtain the following system of linear equations:

$$\sum_{i=0}^{3} c_i^{(2)} \alpha_i^{(2)}\left(x_k^{(2)}\right) = f\left(x_k^{(2)}\right), k = 1, 2, 3, 4$$

By solving this system for the unknown coefficients $C_i^{(2)}, i = 0, 1, 2, 3$ that produces

$$\left. \begin{array}{l} c_0^{(2)} = 0.6366197, c_1^{(2)} = -0.3183099 \\[2mm] c_2^{(2)} = 2.279989 \times 10^{-8}, c_3^{(2)} = -7.819254 \times 10^{-9} \end{array} \right\}$$

(3.1.14)

From (3.1.14) we obtain the approximate solution of Equation (3.1.1) in the form

$$\varphi_n(x) \cong \frac{2\sqrt{1-x^2}}{\pi}(1-x)$$

(3.1.15)

Which coincides with the exact solution. The error of approximate solution (3.1.15) of Equation (3.1.1) at n=20 is given by Table 1.

Thirdly, let us consider in detail the case (III), j=3, for n=3. This results in

$$u_{0..i}^{(3)}(x) = \int_{-1}^{1} \sqrt{\frac{1+t}{1-t}} \frac{V_i(t)}{t-x} dt, u_{2..i}^{(3)}(x) = \int_{-1}^{1} \sqrt{\frac{1+t}{1-t}} \frac{t^2 V_i(t)}{t-x} dt, -1 < t < 1,$$

(3.1.16)

$$\gamma_{0,i}^{(3)} = \int_{-1}^{1} \sqrt{\frac{1+t}{1-t}} V_i(t) dt, \quad \gamma_{3,i}^{(3)} = \int_{-1}^{1} \sqrt{\frac{1+t}{1-t}} t^3 V_i(t) dt,$$

(3.1.17)

By applying the following relations

$$\int_{-1}^{1} \sqrt{\frac{1+t}{1-t}} V_i(t) V_j(t) dt = \begin{cases} 0 & i \neq j \\ \pi & i = j \end{cases}$$

(3.1.18)

$$\int_{-1}^{1} \sqrt{\frac{1+t}{1-t}} \frac{V_i(t)}{t-x} dt = \pi W_i(x)$$

(3.1.19)

It is easy to estimate the values $u_{0,i}^{(1)}, u_{2,i}^{(1)}, \gamma_{0,i}^{(1)}$ and $\gamma_{3,i}^{(1)}$.

Table 1: Illustrates errors of approximate solutions of Equation (3.1.1) in Cases (I)-(IV) at n = 20

x	error (j = 1)	error (j = 2)	error (j = 3)	error (j = 4)
−9.500000E-01	0.000000E+00	0.000000E+00	5.960464E-08	5.960464E-08
−9.000000E-01	0.000000E+00	0.000000E+00	1.192093E-07	1.192093E-07
−7.0000000E-01	0.000000E+00	0.000000E+00	1.192093E-07	1.192093E-07
−5.000000E-01	5.960464-08	5.960464E-08	1.788139E-07	1.788139E-07
−3.000000E-01	0.000000E+00	5.960464E-08	1.788139E-07	1.788139E-07
−1.000000E-01	5.960464E-08	5.960464E-08	1.192093E-07	1.192093E-07
0.000000E+00	5.960464E-08	5.960464E-08	1.192093E-07	1.192093E-07
1.000000E-01	1.192093E-07	5.960464E-08	5.960464E-08	5.960464E-08
3.000000E-01	8.940697E-08	8.940697E-08	8.940697E-08	8.940697E-08
5.000000E-01	8.940697E-08	8.940697E-08	5.960464E-08	5.960464E-08
7.000000E-01	1.043081E-07	7.450581E-08	1.490116E-08	1.490116E-08
9.000000E-01	9.313226E-08	5.029142E-08	4.656613E-08	1.303852E-08
9.500000E-01	5.774200E-08	3.632158E-08	6.705523E-08	4.656613E-09

From the relations (2.4) and (3.1.16)-(3.1.19) we get

$$\alpha_i^{(3)}(t) = \begin{cases} \pi\left(2x^2 + 2x + \dfrac{7}{8}\right) & i = 0 \\[2mm] \pi\left(2x^3 + 3x^2 + x + \dfrac{7}{8}\right) & i = 1 \\[2mm] \pi\left(4x^4 + 6x^3 + x^2 - x + \dfrac{1}{8}\right) & i = 2 \\[2mm] \pi\left(8x^5 + 12x^4 - 5x^2 + x + \dfrac{1}{8}\right) & i = 3 \end{cases} \qquad (3.1.20)$$

By choosing the collocation points

$$x_k^3 = \cos\left(\frac{2k\pi}{2(2n+2)}\right), (k = 1, 2, 3, 4)$$

for n=3, we obtain the following system of linear equations:

$$\sum_{i=0}^{3} c_i^{(3)} \alpha_i^{(3)}\left(x_k^{(3)}\right) = f\left(x_k^{(3)}\right), k = 1, 2, 3, 4$$

By solving this system for the unknown coefficients $C_i^{(3)}, i = 0, 1, 2, 3$ that produces

$$\left.\begin{array}{l} c_0^{(3)} = 0.3183098, c_1^{(3)} = -0.4774647, \\ c_2^{(3)} = 0.1591549, c_3^{(3)} = 1.330901 \times 10^{-8} \end{array}\right\}$$

$$(3.1.21)$$

From (3.1.21) we obtain the approximate solution of Equation (3.1.1) in the form of

$$\varphi_n(x) \cong \frac{2}{\pi}\sqrt{\frac{1+x}{1-x}}\left(x^2 - 2x + 1\right)$$

$$(3.1.22)$$

Which coincides with the exact solution. The error of approximate solution (3.1.22) of Equation (3.1.1) at n=20 is given by Table 1.

Fourthly, In case (IV), j=4, for n=3. This results in

$$u_{0,i}^{(4)}(x) = \int_{-1}^{1}\sqrt{\frac{1-t}{1+t}}\frac{W_i(t)}{t-x}dt, u_{2,i}^{(4)}(x) = \int_{-1}^{1}\sqrt{\frac{1-t}{1+t}}\frac{t^2 W_i(t)}{t-x}dt, -1$$

$$(3.1.23)$$

$$\gamma_{0,i}^{(4)} = \int_{-1}^{1} \sqrt{\frac{1-t}{1+t}} W_i(t)\,dt, \gamma_{3,i}^{(4)} = \int_{-1}^{1} \sqrt{\frac{1-t}{1+t}} t^3 W_i(t)\,dt,$$

(3.1.24)

By applying the relations

$$\int_{-1}^{1} \sqrt{\frac{1-t}{1+t}} W_i(t) W_j(t)\,dt = \begin{cases} 0 & i \neq j \\ \pi & i = j \end{cases}$$

(3.1.25)

$$\int_{-1}^{1} \sqrt{\frac{1-t}{1+t}} \frac{W_i(t)}{t-x}\,dt = -\pi V_i(x)$$

(3.1.26)

It is easy to estimate the values $u_{0,i}^{(4)}, u_{2,i}^{(4)}, \gamma_{0,i}^{(4)}$ and.

From the relations (2.5) and (3.1.23)-(3.1.26) we get

$$\alpha_i^{(4)}(x) = \begin{cases} \dfrac{-7\pi}{8}; & i = 0 \\[2mm] -\pi\left(2x^3 + x^2 - x - \dfrac{7}{8}\right); & i = 1 \\[2mm] -\pi\left(4x^4 + 2x^3 - 3x^2 - x + \dfrac{1}{8}\right); & i = 2 \\[2mm] -\pi\left(8x^5 + 4x^4 - 8x^3 - 3x^2 + x - \dfrac{1}{8}\right); & i = 3 \end{cases}$$

(3.1.27)

By choosing the collocation points

$$x_k^{(4)} = \cos\left(\frac{(2k-1)\pi}{(2n+3)}\right), (k = 1, 2, 3, 4)$$

for n=3 we obtain the following system of linear equations :

$$\sum_{i=0}^{3} c_i^{(4)} \alpha_i^{(4)} \left(x_k^{(4)} \right) = f \left(x_k^{(4)} \right), k = 1, 2, 3, 4$$

By solving this system for the unknown coefficients $C_i^{(4)}, i = 0, 1, 2, 3$ that produces

$$\left. \begin{array}{l} c_0^{(4)} = 0.3183098, c_1^{(4)} = 0.1591549 \\ c_2^{(4)} = -0.1591549, c_3^{(4)} = 2.358931 \times 10^{-8} \end{array} \right\}$$

(3.1.28)

From (3.1.28) we obtain the approximate solution of Equation (3.1.1) in the form of

$$\varphi_n \left(x \right) \cong \frac{-2}{\pi} \sqrt{\frac{1-x}{1+x}} \left(x^2 - 1 \right)$$

(3.1.29)

which coincides with the exact solution. The error of approximate solution (3.2.29) of Equation (3.2.1) at n=20 is given by Table 1.

Example 2:

Consider the following singular integral equation

$$\int_{-1}^{1} \frac{\varphi(t)}{t-x} dt + \int_{-1}^{1} \left(x^3 + xt^2 \right) \varphi(t) dt = x^3 + x$$

(3.2.1)

which corresponds with k(t,x)=1 and L(t,x)=x³+xt². So one gets

$$k_0 \left(x \right) = 1, k_p \left(x \right) = 0, \left(p > 0 \right)$$

$$L_0 \left(x \right) = x^3, L_1 \left(x \right) = 0, L_2 \left(x \right) = x, L_q \left(x \right) = 0 \left(q > 2 \right)$$

Hence we find that relation (2.8) produces

$$\sum_{i=0}^{n} c_i^{(j)} \alpha_i^{(j)}(x) = x^3 + x, -1 < x < 1, (j = 1, 2, 3, 4)$$

Thus (2.9) gives

$$\alpha_i^{(j)}(x) = u_{0,i}^{(j)}(x) + x^3 \gamma_{0,i}^{(j)} + x \gamma_{2,i}^{(j)}, (j = 1, 2, 3, 4), (i = 0, 1, 2, \cdots)$$

Firstly, let us consider in detail the case (I), j=1, for n=3. This results in

$$\gamma_{2,i}^{(1)} = \int_{-1}^{1} \frac{t^2 T_i(t)}{\sqrt{1-t^2}} dt,$$

$$(3.2.2)$$

From the relations (3.1.2)-(3.1.5) and (3.2.2) we obtain

$$\alpha_i^{(1)}(x) = \begin{cases} \pi\left(x^3 + x/2\right) & i = 0 \\ \pi & i = 1 \\ 9\pi x/4 & i = 2 \\ \pi\left(4x^2 - 1\right) & i = 3 \end{cases}$$

$$(3.2.3)$$

By choosing the collocation points

$$x_k = \cos\left(\frac{(2k-1)\pi}{(2n+2)}\right), (k = 1, 2, 3, 4)$$

for n=3 we obtain the following system of linear equations:

$$\sum_{i=0}^{3} c_i^{(1)} \alpha_i^{(1)}(x_k) = f(x_k), k = 1, 2, 3, 4$$

By solving this system for the unknown coefficients $C_i^{(1)}, i = 0,1,2,3$ that produces

$$
\left.\begin{array}{l}
c_0^{(1)} = 0.3183098, c_1^{(1)} = 1.090772 \times 10^{-8} \\
c_2^{(1)} = 0.07073557, c_3^{(1)} = 1.830649 \times 10^{-8}
\end{array}\right\}
$$

(3.2.4)

From (3.2.4) we obtain the approximate solution of Equation (3.2.1) in the form of

$$
\varphi_n(x) \cong \frac{1}{9\pi\sqrt{1-x^2}}\left(7 + 4x^2\right)
$$

(3.2.5)

which coincides with the exact solution. The error of approximate solution (3.2.5) of equation (3.2.1) at n=20 is given by Table 2.

Secondly, let us consider in detail the case (II), j=2, for n=3. This results in

$$
\gamma_{2,i}^{(2)} = \int_{-1}^{1} \sqrt{1-t^2}\, t^2 U_i(t)\, dt,
$$

(3.2.6)

By applying the relations (3.1.9)-(3.1.12) and (3.2.6) we get

$$
\alpha_i^{(2)}(x) = \begin{cases}
\dfrac{\pi}{2}\left(x^3 - \dfrac{7x}{4}\right) & i = 0 \\[2mm]
-\pi\left(2x^2 - 1\right) & i = 1 \\[2mm]
-\pi\left(4x^3 - \dfrac{25x}{8}\right) & i = 2 \\[2mm]
-\pi\left(8x^4 - 8x^2 + 1\right) & i = 3
\end{cases}
$$

(3.2.7)

By choosing the collocation points

$$x_k^{(2)} = \cos\left(\frac{(2k-1)\pi}{(2n+2)}\right), (k = 1,2,3,4)$$

for n=3, we obtain the following

Table 2: Illustrates errors of approximate solutions of Equation (3.2.1) in Case (I), Case (II) and Case (IV) respectively at n = 20

x	Error (j = 1)	Error (j = 2)	Error (j = 3)
−9.500000E-01	0.000000E+00	0.000000E+00	0.000000E+00
−9.000000E-01	5.960464E-08	5.960464E-08	0.000000E+00
−7.000000E-01	8.940697E-08	1.192093E-07	5.960464E-08
−5.000000E-01	8.940697E-08	1.192093E-07	1.192093E-07
−3.000000E-01	8.940697E-08	1.788139E-07	1.192093E-07
−1.000000E-01	1.192093E-07	1.788139E-07	1.788139E-07
0.000000E+00	1.043081E-07	1.788139E-07	1.788139E-07
1.000000E-01	1.192093E-07	1.788139E-07	1.192093E-07
3.000000E-01	8. 940697E-08	1.788139E-07	5.960464E-08
5.000000E-01	8.940697E-08	1.192093E-07	1.192093E-07
7.000000E-01	8.940697E-08	1.192093E-07	0.00000E+00
9.000000E-01	5.9604641E-08	5.960464E-08	5.960464E-08
9.500000E-01	0.000000E+00	0.000000E+00	0.000000E+00

system of linear equations:

$$\sum_{i=0}^{3} c_i^{(2)} \alpha_i^{(2)}\left(x_k^{(2)}\right) = f\left(x_k^{(2)}\right), k = 1,2,3,4$$

By solving this system for the unknown coefficients $C_i^{(2)}, i = 0,1,2,3$ that produces

$$c_0^{(2)} = -1.170559, c_1^{(2)} = -1.331665 \times 10^{-9}$$
$$c_2^{(2)} = -0.2258973, c_3^{(2)} = -1.644008 \times 10^{-8}$$

$$(3.2.8)$$

From (3.2.8) we obtain the approximate solution of Equation (3.2.1) in the form of

$$\varphi_n(x) \cong \frac{-\sqrt{1-x^2}}{31\pi}\left[92 + 88x^2\right]$$

$$(3.2.9)$$

which coincides with the exact solution. The error of approximate solution (3.2.9) of Equation (3.2.1) at n=20 is given by Table 2.

Thirdly, In case (IV), j=4, for n=3. This results in

$$\gamma_{2,i}^{(4)} = \int_{-1}^{1} \sqrt{\frac{1-t}{1+t}} t^2 W_i(t)\, dt,$$

$$(3.2.10)$$

By applying the relations (3.1.23)-(3.1.26) and (3.2.10) we get

$$\alpha_i^{(4)}(x) = \begin{cases} \pi\left(x^3 + \dfrac{x}{2} - 1\right) & i = 0 \\[2mm] -\pi\left(\dfrac{9}{4}x - 1\right) & i = 1 \\[2mm] -\pi\left(4x^2 - \dfrac{9x}{4} - 1\right) & i = 2 \\[2mm] -\pi\left(8x^3 - 4x^2 - 4x + 1\right) & i = 3 \end{cases}$$

$$(3.2.11)$$

By choosing the collocation points

$$x_k^{(4)} = \cos\left(\frac{(2k-1)\pi}{(2n+3)}\right), (k = 1,2,3,4)$$

for n=3 we obtain the following system of linear equations:

$$\sum_{i=0}^{3} c_i^{(4)} \alpha_i^{(4)}\left(x_k^{(4)}\right) = f\left(x_k^{(4)}\right), k = 1,2,3,4$$

By solving this system for the unknown coefficients $C_i^{(4)}, i = 0,1,2,3$ that produces

$$c_0^{(4)} = c_1^{(4)} = -0.5852794, c_2^{(4)} = c_3^{(4)} = -0.1129487 \tag{3.2.12}$$

From (3.2.12) we obtain the approximate solution of Equation (3.2.1) in the form of

$$\varphi_n(x) \cong \frac{-1}{31\pi}\sqrt{\frac{1-x}{1+x}}(1+x)(92+88x^2) \tag{3.2.13}$$

which coincides with the exact solution. The error of approximate solution (3.2.13) of Equation (3.2.1) at n=20 is given by Table 2.

Similarly, doing the same operations as we did for Case (I), Case (II) and Case (IV), one can solve for Case (III).

Example 3:

Consider the following singular integral equation

$$\int_{-1}^{1} \frac{\varphi(t)}{t-x} dt + \int_{-1}^{1} \left(x^2 + t^2\right)\varphi(t) dt = \frac{-3}{2}x^2 + 2x, \tag{3.3.1}$$

which corresponds with k(t, x)=1 and L(t, x)=x²+t². So, one gets

$$k_0(x) = 1, k_p(x) = 0, (p > 0)$$

$$L_0(x) = x^2, L_1(x) = 0, L_2(x) = 1, L_q(x) = 0 (q > 2)$$

Hence the relation (2.8) produces

$$\sum_{i=0}^{n} c_i^{(j)} \alpha_i^{(j)}(x) = \frac{-3}{2} x^2 + 2x, -1 < x < 1, j = 1, 2, 3, 4$$

$$(3.3.2)$$

where (2.9) gives

$$\alpha_i^{(j)}(x) = u_{0,i}^{(j)}(x) + x^2 \gamma_{0,i}^{(j)} + \gamma_{2,i}^{(j)}, (j = 1, 2, 3, 4), (i = 0, 1, 2, \cdots)$$

Firstly, let us consider in detail the case (II), j=2, for n=3. From (3.1.9)-(3.1.12) and (3.2.6) we get

$$\alpha_i^{(2)}(x) = \begin{cases} \dfrac{\pi}{8}\left(4x^2 - 8x + 1\right) & i = 0 \\[2mm] -\pi\left(2x^2 - 1\right) & i = 1 \\[2mm] \dfrac{-\pi}{8}\left(32x^3 - 24x - 1\right) & i = 2 \\[2mm] -\pi\left(8x^4 - 8x^2 + 1\right) & i = 3 \end{cases}$$

$$(3.3.3)$$

By solving the system (3.3.2), at the collocation points

$$x_k^{(2)} = \cos\left(\frac{(2k-1)\pi}{(2n+2)}\right), (k = 1, 2, 3, 4)$$

for the unknown coefficients $C_i^{(2)}, i = 0, 1, 2, 3$ we obtain

$$c_0^{(2)} = -0.6366197, c_1^{(2)} = 0.07957754$$
$$c_2^{(2)} = 1.746461 \times 10^{-8}, c_3^{(2)} = 1.827517 \times 10^{-8} \Big\}$$

$$(3.3.4)$$

So the approximate solution of Equation (3.3.1) is given by

$$\varphi_n(x) \cong \frac{-\sqrt{1-x^2}}{2\pi}(4-x),$$

$$(3.3.5)$$

which coincides with the exact solution, the error of the approximate solution (3.3.5) of Equation (3.3.1) at n=20 is given by Table 3.

Secondly, in case (III), j=3, for n=3. This results in

$$\gamma_{2,i}^{(3)} = \int_{-1}^{1} \sqrt{\frac{1+t}{1-t}} t^2 V_i(t) dt,$$

$$(3.3.6)$$

From (3.1.16)-(3.1.19) and (3.3.6) we get

$$\alpha_i^{(3)}(x) = \begin{cases} \left| \pi\left(x^2 + \dfrac{3}{2}\right) \right| & i = 0 \\[2mm] \pi\left(2x + \dfrac{5}{4}\right) & i = 1 \\[2mm] \pi\left(4x^2 + 2x - \dfrac{3}{4}\right) & i = 2 \\[2mm] \left| \pi\left(8x^3 + 4x^2 - 4x + 1\right) \right| & i = 3 \end{cases}$$

$$(3.3.7)$$

By solving the system (3.3.2), at the collocation points

$$x_k^{(3)} = \cos\left(\frac{(2k-1)\pi}{(2n+3)}\right), (k = 1, 2, 3, 4)$$

for the unknown coefficients $C_i^{(3)}, i = 0, 1, 2, 3$ we obtain

$$\left.\begin{array}{l} c_0^{(3)} = -0.3183099, c_1^{(3)} = 0.3580987, \\ c_2^{(3)} = -0.03978872, c_3^{(3)} = -8.155105 \times 10^{-9} \end{array}\right\} \qquad (3.3.8)$$

Hence, the approximate solution of Equation (3.3.1) is given by

$$\varphi_n(x) \cong \frac{-1}{2\pi} \sqrt{\frac{1+x}{1-x}} \left(x^2 - 5x + 4\right) \qquad (3.3.9)$$

which coincides with the exact solution, the error of the approximate solution (3.3.9) of Equation (3.3.1) at n=20 is given by Table 3.

Table 3: Illustrates errors of approximate solutions of Equation (3.3.1) in Case (II) and Case (III) at n = 20

x	Error (j = 2)	Error (3)
−9.500000E-01	2.980232E-08	2.980232E-08
−9.000000E-01	2.980232E-08	5.960464E-08
−7.000000E-01	0.000000E+00	5.960464E-08
−5.000000E-01	0.000000E+00	1.192093E-07
−3.000000E-01	0.000000E+00	1.192093E-07
−1.000000E-01	5.960464E-08	1.192093E-07
0.000000E+00	5.960464E-08	1.192093E-07
1.000000E-01	5.960464E-08	1.192093E-07
3.000000E-01	1.192093E-07	1.192093E-07
5.000000E-01	1.192093E-07	8.940697E-08
7.000000E-01	1.192093E-07	0.000000E+00
9.000000E-01	8.940697E-08	1.788139E-07
9.500000E-01	5.960464E-08	3.278255E-07

Similarly, doing the same operations as we did for Case (II) and Case (III), one can solve for Case (I) and Case (IV).

CONCLUSIONS

Numerical results (Tables 1-3) show that the errors of approximate solutions of Examples 1-3 in different Cases with small value of n are very small. These show that the methods developed are very accurate and in fact for a linear function give the exact solution.

REFERENCES

1. Chakrabarti, A. (1989) Solution of Two Singular Integral Equations Arising in Water Wave Problems. ZAMM, 69, 457-459. http://dx.doi.org/10.1002/zamm.19890691209

2. Ladopoulous, E.G. (2000) Singular Integral Equations Linear and Non-Linear Theory and Its Applications in Science and Engineering. Springer, Berlin.

3. Ladopoulous, E.G. (1987) On the Solution of the Two-Dimensional Problem of a Plane Crack of Arbitrary Shape in an Anisotropic Material. Engineering Fracture Mechanics, 28, 187-195. http://dx.doi.org/10.1016/0013-7944(87)90212-8

4. Zabreyko, P.P. (1975) Integral Equations—A Reference Text. Noordhoff, Leyden. http://dx.doi.org/10.1007/978-94-010-1909-5

5. Prossdorf, S. (1977) On Approximate Methods for the Solution of One-Dimensional Singular Integral Equations. Applicable Analysis, 7, 259-270.

6. Zisis, V.A. and Ladopoulos, E.G. (1989) Singular Integral Approximations in Hilbert Spaces for Elastic Stress Analysis in a Circular Ring with Curvilinear Cracks. Indus. Math., 39, 113-134.

7. Chakrabarti, A. and Berghe, V.G. (2004) Approximate Solution of Singular Integral Equations. Applied Mathematics Letters, 17, 553-559. http://dx.doi.org/10.1016/S0893-9659(04)90125-5

8. Abdou, M.A. and Naser, A.A. (2003) On the Numerical Treatment of the Singular Integral Equation of the Second Kind. Applied Mathematics and Computation, 146, 373-380.http://dx.doi.org/10.1016/S0096-3003(02)00587-8

9. Abdulkawi, M., Eshkuvatov, Z.K. and Nik Long, N.M.A. (2009) A Note on the Numerical Solution of Singular Integral Equations of Cauchy Type. International Journal of Applied Mathematics and Computer Science, 5, 90-93.

10. Eshkuvatov, Z.K., Nik Long, N.M.A. and Abdulkawi, M. (2009) Approximate Solution of Singular Integral Equations of the First Kind with Cauchy Kernel. Applied Mathematics Letters, 22, 651-657. http://dx.doi.org/10.1016/j.aml.2008.08.001

11. Gakhov, F.D. (1966) Boundary Value Problems. Addison-Wesley, Boston.

12. Martin, P.A. and Rizzo, F.J. (1989) On Boundary Integral Equations for Crack Problems. Proceedings of the Royal Society A, 421, 341-345.http://dx.doi.org/10.1098/rspa.1989.0014

13. Sheshko, M. (2003) Singular Integral Equations with Cauchy and Hilbert Kernels and Their Approximated Solutions. The Learned Society of the Catholic University of Lublin, Lublin. (in Russian)

14. Muskhelishvili, N.I. (1977) Singular Integral Equations. Noordhoff International Publishing, Leyden. http://dx.doi.org/10.1007/978-94-009-9994-7

15. Lifanov, I.K. (1996) Singular Integral Equation and Discrete Vortices. VSP, Leiden.

16. Kyth, K.P. and Schaferkotter, R.M. (2005) Handbook of Computational Methods for Integration. Chapman & Hall/ CRC Press, London.

CITATION

Dardery, S. and Allan, M. (2014) Chebyshev Polynomials for Solving a Class of Singular Integral Equations. Applied Mathematics, 5, 753-764. doi: 10.4236/am.2014.54072.

Index